Energy from Biomass

Solar Energy R&D in the European Community

Series E:

Energy from Biomass

Volume 2

Publication arrangements: D. NICOLAY

Solar Energy R&D
in the European Community

Series E Volume 2

Energy from Biomass

Proceedings of the Workshop on
Biomass Pilot Projects on
Methanol Production and Algae,
held in Brussels, 22 October 1981

edited by

W. PALZ and G. GRASSI
Commission of the European Communities

D. REIDEL PUBLISHING COMPANY

Dordrecht, Holland / Boston, U.S.A. / London, England

for the Commission of the European Communities

Library of Congress Cataloging in Publication Data

CIP

Main entry under title:

Energy from biomass

(Solar energy R&D in the European community. Series E ; v. 2)
1. Biomass energy–Congresses. 2. Methanol as fuel–Congresses.
I. Palz, Wolfgang. II. Grassi, G., 1929- III. Commission of the European Communities. IV. Series.
TP360.B594 338.4'7665776 81-19977
ISBN 90-277-1370-7 AACR2

Organization of the workshop by
Commission of the European Communities
Directorate-General Research, Science and Development, Brussels

Publication arrangements by
Commission of the European Communities
Directorate-General Information Market and Innovation, Luxembourg

EUR 7667

Published by D. Reidel Publishing Company
P.O. Box 17, 3300 AA Dordrecht, Holland

Sold and distributed in the U.S.A. and Canada
by Kluwer Boston Inc.,
190 Old Derby Street, Hingham, MA 02043, U.S.A.

In all other countries, sold and distributed
by Kluwer Academic Publishers Group,
P.O. Box 322, 3300 AH Dordrecht, Holland

D. Reidel Publishing Company is a member of the Kluwer Group

Printed in The Netherlands.

PREFACE

This book comes as part of a new series on Solar Energy R+D, including Biomass which is carried out by the European Community. The Commission of the European Communities' Directorate General (XII) for Science, Research and Development is currently implementing, on a cost-sharing basis, a solar energy R+D programme through contracts with European industry, research institutions and universities. This programme includes a very strong activity on Biomass. Besides general R+D work on all aspects of Biomass growth and utilization which is reported elsewhere in this series, the Commission is currently starting a new activity on Pilot Plants based on the use of Biomass for energy purposes, and in particular on methanol production from wood.

The Commission considers that the subject of methanol production from wood offers important prospects for application within the European Community and in other parts of the world, in particular some of the developing countries. The state of art in Europe in this field is still considered to be very high as a result of related work which was performed in Europe during World War II and the time before.

The present book starts with a review paper on the state of art as it appears presently in Europe. Furthermore, the book reports on the workshop which has been held at the Commission in Brussels on 22 October 1981 and at which 9 proposals on the subject of methanol production from wood were presented. The papers give a comprehensive overview of the various technologies presently considered in Europe on the subject of methanol production from wood. In the course of the Commission's activity, four of the nine proposals submitted will actually be built; one in France, one in the U.K., one in Germany and one in Italy. They will all be completed and commissioned in the second half of 1983.

This book also reports on a few proposals for pre-pilot projects on Algae, in particular on the species Botryococcus braunii which offers considerable hope in producing petrol directly from the sun.

I hope that this book will attract the attention it deserves from the biomass community and can contribute to the distribution of information in this important sector.

W. PALZ

CONTENTS

Methanol from Wood; a review related to the proposals submitted to the Commission of the European Communities.

A.A.C.M. Beenackers
and
W.P.M. van Swaaij

Twente University of Technology
P.O. Box 217
7500 AE ENSCHEDE
The Netherlands

Abstract

Long term potential of methanol from wood in the EEC is briefly discussed. The gasification technology involved in the various proposals for participation in the C.E.C. project E, "Energy from Biomass", pilot project, "Synthetic Fuel from Wood" has been classified along the possible routes. These proposals include oxygen gasification, steam gasification and gasification with chemically bound oxygen produced in a separate reactor from air and a reacting solid. Reactors suggested are moving bed gasifiers (both co- and cross current) fluidized bed gasifiers (both single- and multiple beds), a "fast" or circulating fluidized bed and a powder flame gasifier. Some general strong points and problem areas of the various routes are pointed out. No rating of the different proposals is given in this overview to allow for an open discussion.

Introduction

On request of the Directorate General for Research, Science and Development of the Commission of the European Communities, we analysed proposals for participation in the C.E.C. program Project E, "Energy from Biomass", pilot project "Synthetic Fuel from Wood". In the course of our investigations we received much additional information from the proposers who were found to be very cooperative. Because often the additional information was considered as confidential and to give way to an open discussion, this review is not meant to be an in depth analysis of the characteristics, merits and risks of the projects proposed (this was presented elsewhere) but rather a classification scheme providing some general background information. We will only point out some general strong and weak points of the various processes as far as can be derived from open literature, including the summaries presented by the proposers as published in this book.
Additionally we'll present a personal view on the opportunities for methanol production from biomass in Europe and give a short description of the technologies involved.

An outlook on methanol from biomass

Biomass in general and wood in particular have regained increasing interest as alternative feedstock for energy carriers over the past decade.
Its advantages and disadvantages are summarized in Table I.

Table I.

Wood as an energy source; advantages and problems.

Advantages	Disadvantages
- continuous flow source (solar energy)	- it is a solid - high moisture content
- clean source of energy, little sulfur and heavy metals	- low bulk density
- in steady state no net CO_2 production	- high transportation costs
- economic incentive for renewed interest in forestry	- high collection costs
- decentralized production in many areas possible, requiring little attendance	- large areas necessary (low energy production density)
	- competition with food production might be a problem.

Possible energy carriers resulting from the principal biomass conversion processes are presented in Table II.
Recent reviews [1, 2] indicate that production of methanol probably will be the more prospective large scale option, relative to SNG, ethanol and oxyhydrocarbon production. If gasoline production from biomass will ever become feasible the indirect synthesis via methanol presently looks more prospective than the Fisher Tropsch alternative [1].

Table II.

Energy carriers from conversion of biomass.

- heat
- steam
- electricity
- gaseous fuels: - low Joule gas
  - medium Joule gas
  - high Joule gas
- liquid fuels:
  - methanol (from synthesis gas)
  - ethanol (fermentation)
  - gasoline (either from synthesis gas or via methanol)
  - oxyhydrocarbons (by pyrolysis or liquefaction).

Some of the disadvantages of wood as a feedstock for fuels become increasingly important with increasing scale of application (low bulk density, high transportation costs, high collection costs, low energy production density). Therefore, the viability of large scale methanol production from wood is less obvious than the typical small and intermediate scale wood conversion processes (heat, steam and local electricity generation by low or medium Joule gas from gasification).
The applications in the latter category have been, and still are, developed under the previous and present C.E.C. "Energy from Biomass" projects.
Nowadays, methanol is mainly produced by steam reforming of natural gas and sometimes by oxygen gasification of petroleum residues. Production via gasification of caol is still insignificant but is rapidly gaining interest.
Methanol plants presently have capacities typically in the order of 10^6 kg/day [3]; also feasibility studies on methanol production from biomass published so far are almost exclusively restricted to the capacity range 0.6 - 2.5 10^6 kg methanol/day [4]. On that production level however, biomass gasification is not expected to find wide application [4, 5]. Restrictions are the availability of coal as a large scale alternative and the low energy production density of biomass leading to increasing transportation costs with increasing capacity (typically 10^6 kg methanol/day asks for 1500 km^2 of average forest land).
It is generally agreed that wide application of methanol from biomass will depend on whether or not it can be produced economically at the 10^5 kg/day level [6, 7]. At this scale, feasibility is still uncertain and the main problem therefore is down scaling the methanol from biomass process to a viable intermediate scale.
Figure 1. reviews projected methanol production costs from both biomass and conventional recources including coal (costs extrapolated to 1980 dollars) [4, 1]
It shows the economy of scale of presently available technology and the viability of coal as a feedstock for very large scale methanol production plants in the mining areas. It must be kept in mind however that the figures for coal based plants relate to the US where coal is mined much cheaper than in the EEC. Figure 1. also shows a large scatter in data, especially for methanol from biomass. In part, this is due to the

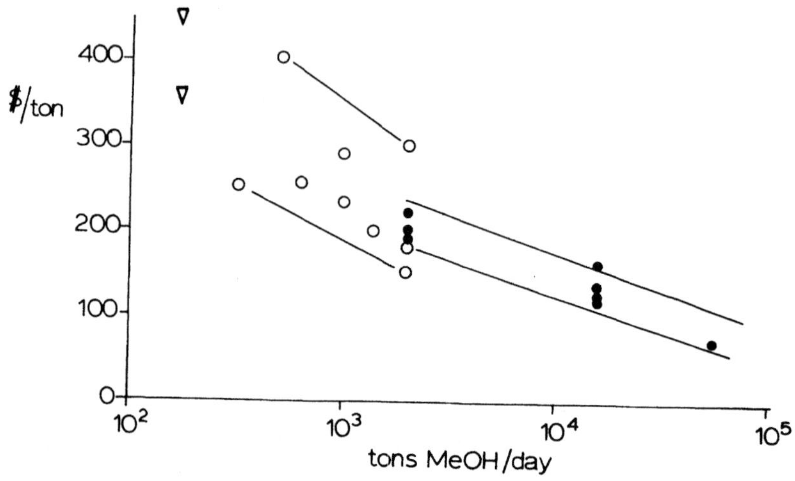

Figure 1. Estimates of methanol production costs (1980 $/ton)
• conventional including coal based, technology SERI [4];
o from wood, SERI [4];
∇ from wood, Ader et. al. [1].

uncertainty in overall process efficiency that can be obtained (see Table III); in part, it also depends on the highly site specific cost price of the feedstock wood which is a substantial cost factor (up to 40 % [1]).

Table III.

Predicted energy efficiency for methanol from wood.

MITRE [2]	45 - 55 %
SAI [8]/SERI [4]	30 - 47 %
Ader, Bridgwater and Hatt [1]:	
low temp. pyrolysis	35 %
high temp. pyrolysis	41 %
oxygen gasification	48 %
steam gasification	55 %

The efficiency depends rather strongly on the product quality of the raw gas and the corresponding need of steam reforming. That is the main reason why pyrolysis processes are not attractive and expensive oxygen gasification may turn out to be competitive with steam gasification. It is also the reason why methanol from coal often has a somewhat better energy efficiency than methanol from wood.

From the above considerations we conclude that with present technology and if coal prices do not rise relatively excessively, methanol from wood often will be slightly more expensive than methanol from coal. Nevertheless it may find its way thanks to political incentives as:
- energy independence,
- regional development,
- environmental considerations, and
- forestry development,

provided that funds are made available and a policy is adopted to debottle-neck the expected constraints for a "business as usual" development. The C.E.C. "Methanol from Wood" program is intended to contribute to this.

One additional remark on the economics of scale can be made here: forestry nearby an existing conventional or coal based methanol plant may produce synthesis gas and sell it to the methanol facility, thus benefiting from the favourable economics of large scale methanol production.

Methanol technology.

Manufacturing of methanol from synthesis gas is considered to be a proven technology and accordingly the C.E.C. Call for Tenders "Methanol from Wood" only requires production of a gas that is suitable as a feedstock for methanol production. We will therefore briefly discuss here present technology to define synthesis gas feed specifications and favourable process conditions.

Methanol is produced mainly from synthesis gas according to:

$$CO + 2H_2 \rightleftarrows CH_3OH \qquad \Delta H_{298} = -90.8 \text{ kJ/mol}$$

All methanol capacity built since 1966 has used the low pressure technology (ICI, LURGI) rather than older high pressure technology [19].

Figure 2 shows a simplified scheme.

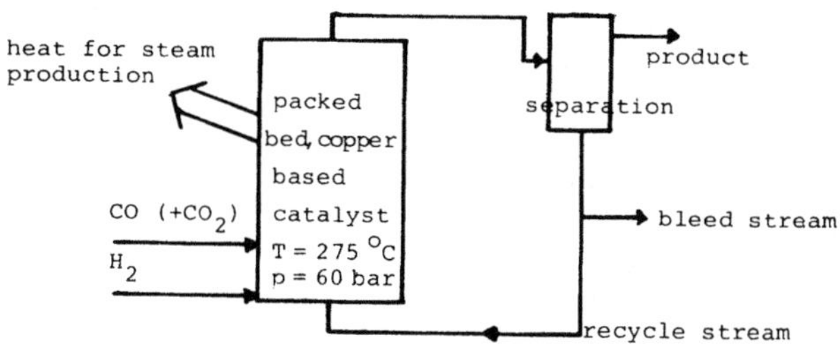

Figure 2. Scheme for low pressure methanol production.

The optimal ratio of H_2/CO in the feedgases should be around 2, though, some CO_2 can be tolerated:

$$CO_2 + 3H_2 \rightleftarrows CH_3OH + H_2O \qquad \Delta H = -49.5 \text{ kJ/mol}$$

In practice, an optimum composition is [10]

$H_2/(2CO + 3CO_2) \simeq 1.05$.

The gas should be free of tars and extremely low in H_2S content (< 0.06 ppm [3]) to avoid catalyst poisoning. If other components are present (e.g. CH_4, N_2, etc.), this will increase the bleedstream which reduces the efficiency of the process. In the case of hydrocarbons present (CH_4, etc.) one could also reform these gases to $CO + H_2$ or reform the rest gases and cycle them back to the reactor. This complicates the system however and increases the capital cost.

It is therefore clear that the ideal feedstock for the methanol plant should have a balanced ratio H_2: $(CO + CO_2)$, contain no sulphur and only minor amounts of N_2, CH_4 and other hydrocarbons.

Feedstocks from gasification do not fulfill all these requirements. Sulphur and CO_2 can be removed by well known absorption processes. The right ratio of H_2/CO can be realized by inserting a so called shiftreactor. The slightly exothermic reaction

$$CO + H_2O \rightleftarrows CO_2 + H_2$$

can be shifted towards $CO_2 + H_2$ over a packed bed of catalyst by addition of steam. Usually only relatively minor shifts are necessary and the technique is proven and simple and needs no further discussion here.

A last consideration, affecting efficient gasification, is that the molar gaseous input into the gasifier is substantially smaller than that into the methanol reactor; typically the molar ratio is in the order of 1:2.5. A pressurized gasifier therefore will reduce overall compression costs. This must be balanced against the increased capital investment and often a higher hydrocarbon content in the produced gas, especially with low temperature gasifiers (< 1200 °C). The optimum gasification pressure will probably be somewhere between 10 and 30 bar depending on the type of process

Gasification methods.

Methanol synthesis requires a medium Joule value gas. There are at least four different methods to produce such a gas (Table IV) and the proposals presented in this book cover three of them. For method D no suitable technology has been developed as yet.

Table IV

Gasification methods.

method	proposers
A. gasification with oxygen or oxygen/steam mixtures	Creusot-Loire [13] Cemagref [15] Novelerg [16] Foster Wheeler [17] Lurgi [19]
B. gasification with steam while combusting part of the biomass separately with air to provide heat	U.L.B. [11] IRD-PPC [14] Fritz Werner [18]
C. gasification with chemically bound oxygen produced in a separate reactor from air and a reacting solid	John Brown-Wellman [12]
D. gasification with air and extraction of $CO + H_2$ from the low Joule value gases produced	

Gasification with oxygen (and steam)

A simplified scheme is indicated in Figure 3.
Advantages are the relatively good quality of the product gases containing mainly CO and H_2. The composition depends somewhat on the operation temperature, e.g. at low temperatures formation of methane and tars will occur. A further advantage is its suitability for gasifying wet feedstocks. Problems are the high cost of oxygen and the complexity of safe operation. These disadvantages become specially important in industrial

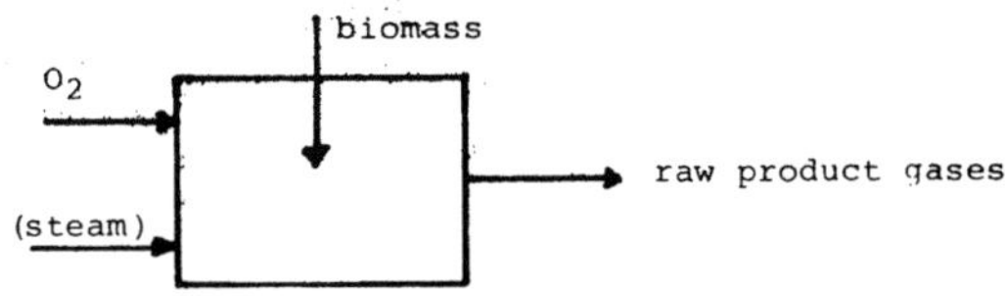

Figure 3. Single stage oxygen (+steam) gasifier.

units of low capacities.
The flexibility of the plant could be increased and the tar and methane content decreased by utilizing a two stage reactor (see figure 4).

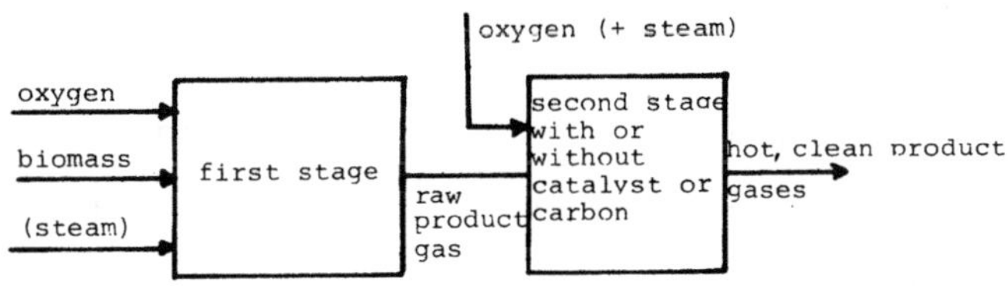

Figure 4. Two stage gasifier with oxygen.

This solution has been adapted in some proposals. The disadvantage is the complexity and increased capital costs.
In one proposal (Novelerg, [16]) the oxygen consumption is minimized by applying electricity as an additional heat source. The feasibility of this method is limited to sites where surplus capacity of very cheap electricity from nuclear or hydropower stations exists. Additional disadvantages possibly are an increased tar and hydrocarbon production and a more complex operation.

Gasification with steam while combusting part of the biomass separately with air to generate heat.

This concept is derived from the classical town gas process in which air and steam were introduced intermittently. In new developments, a heat carrier mixed with partially converted biomass is circulated continuously between two fluid bed vessels (see figure 5). The concept has been indtroduced previously (e.g. by Batelle [20]).
In the combustor the remaining char is combusted, thus heating up the solids with, say, an additional 200 °C. This sensible heat is used in the gasifier to produce synthesis gas via the endothermic reaction

$$C + H_2O \rightleftarrows CO + H_2$$

An advantage is the utilization of air and steam instead of oxygen. Important problems are the complexity of the operation

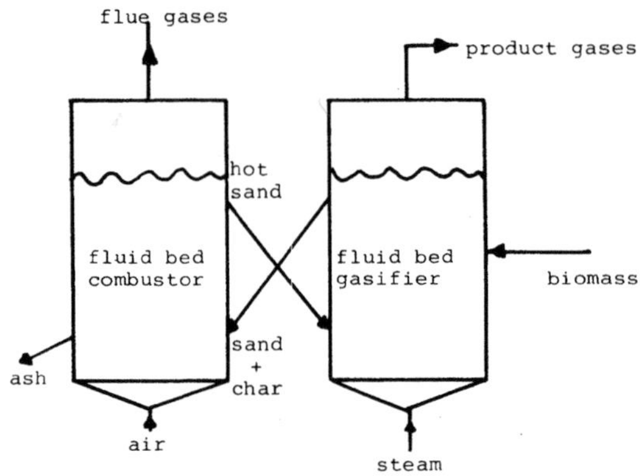

Figure 5. Steam gasification with double fluidized bed process.

and the expected difficulties in pressurizing the plant. Another serious draw back is the low gasification temperature which results in a product gas containing much tar, methane and other hydrocarbons. For methanol production these systems therefore should be provided with a secondary gasification reactor. The reason for the low gasification temperature is the upper temperature limit in the combustor imposed by the need to avoid ash fusion and by the temperature difference between gasifier and oxidiser, needed to allow for sufficient heat transport. In one paper [11], a novel concept is introduced to shorten transport lines and to realize a compact design. In another paper [18] a concentric fluidized bed is proposed with indirect heat transfer from combustor to gasifier by conduction through the wall.

Gasification with chemically bound oxygen (oxygen donor process)

This is a new process not yet applied in any form in practice. Here also a heat carrier is circulated between two vessels. However apart from heat also oxygen is transferred from the combustor/oxydiser to the gasifier (see figure 6).
In the combustor the most important exothermic reaction is

$$CaS + 2O_2 \rightarrow CaSO_4$$

Also a small amount of char will be combusted. All these reactions increase the temperature of the solid.
In the gasifier recycle gas is oxidised

$$CaSO_4 + 4H_2 \rightarrow CaS + 4H_2O$$
$$CaSO_4 + 4CO \rightarrow CaS + 4CO_2$$

The H_2O and CO_2 formed, then react further with the hot biomass and char present in the gasifier to produce H_2 and CO.
Advantages claimed are the utilization of air instead of oxygen and less or no steam consumption in the gasifier.
Problems are the combination of a complex physical system (two fluid beds) and the complex reaction system. Some SO_2 may be emitted from the regenerator (combustor) at low partial oxygen pressures, which is a disadvantage, especially because

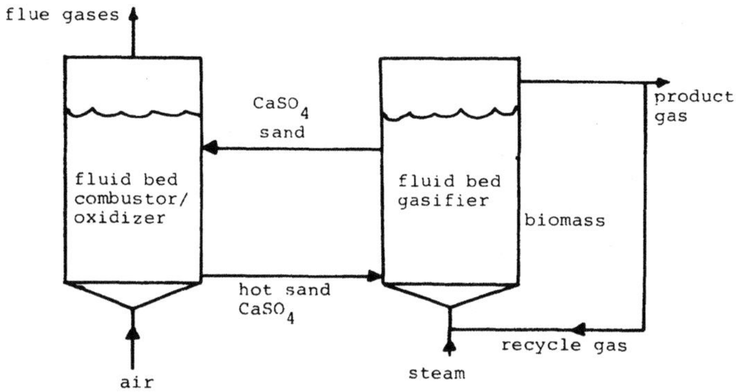

Figure 6. Gasification by Oxygen Donor Process.

biomass itself contains little sulphur.

Reactor type selection.

Quite a number of gasification reactor types are under development or already commercially available. Altogether, six different types are considered in the various proposals submitted to the C.E.C. (Table V). Figure 7 gives an overview with some advantages (+) and disadvantages (-) for each reactor.

Table V

reactor type	proposer
- moving bed, cocurrent	Novelerg [16]
- moving bed, cross current	Cemagref [15] Foster Wheeler [17]
- single fluid bed	Creusot Loire [13]
- multiple fluid beds	ULB [11] John Brown-Wellman [12] IRD-PPC [14] Fritz Werner [18]
- "fast" or circulating fluid bed	Lurgi [19]
- powder flame gasifier	Foster Wheeler [17]

Co-current moving bed.

This reactor has a favourable contacting pattern for pyrolysis product conversion and easy operation performance but is difficult to scale up, specially for irregular shaped feedstocks as biomass often is. In one proposal (Novelerg) this reactor is selected while applying a recycle of pyrolysis gas. This latter fact could make scaling up less problematic but may given rise to permeability problems in the recycle section where tars are present.

CO CURRENT MOVING BED

+ SIMPLE FEEDING
+ LOW TAR
+ LOW METHANE
- DIMENSION FEEDSTOCK
- SCALING UP (?)

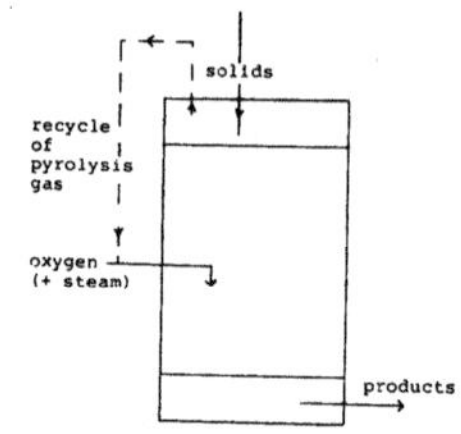

CROSS CURRENT MOVING BED

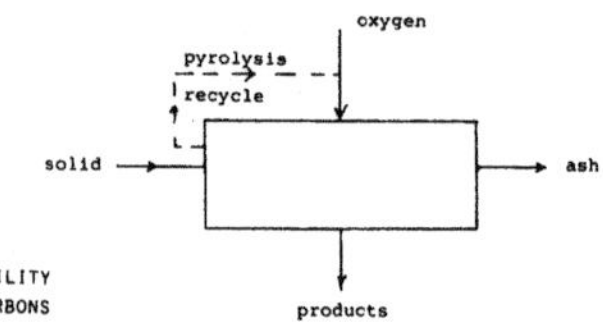

+ BETTER PERMEABILITY
- TARS & HYDROCARBONS

FLUIDIZED BED

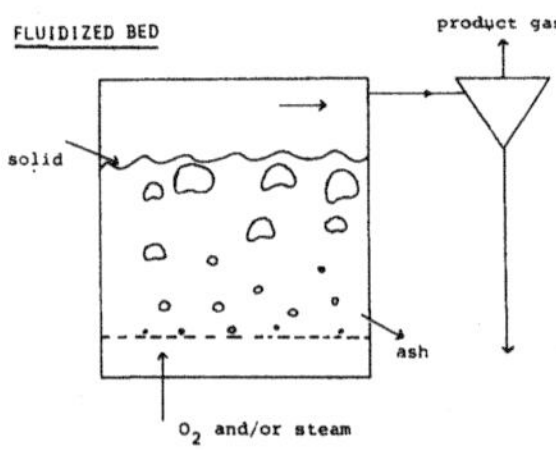

+ OMNIVOROUS AND FLEXIBLE
+ SCALE UP
- LIMITED PYROLYSIS PRODUCTS CONVERSION
- TEMPERATURE LIMITED BY ASH FUSION
- (LIMITED CHAR CONVERSION)
- FEEDSTOCK SIZE LIMITED

DOUBLE FLUIDIZED BEDS

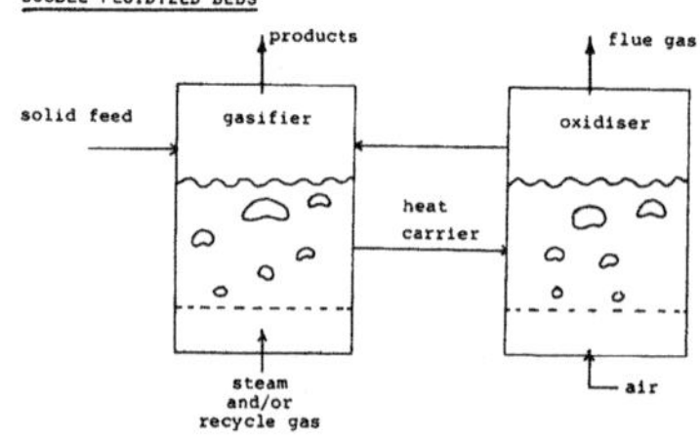

+/- ALL ADVANTAGES AND DISADVANTAGES OF SINGLE FLUIDIZED BEDS
+ NO OXYGEN REQUIRED
- USUALLY MORE TAR AND METHANE DUE TO ASH FUSION TEMP. LIMIT IN OXIDISER AND CONSEQUENTLY LOWER MAXIMUM TEMPERATURE IN GASIFIER
- DIFFICULT TO OPERATE, ESPECIALLY UNDER PRESSURE

FAST OR RECYLING FLUID BED

+ FEEDSTOCK FLEXIBILITY
+ HIGH TEMPERATURE POSSIBLE
- EROSION AND ATTRITION (?)
- CONTROLABILITY WITH BIOMASS

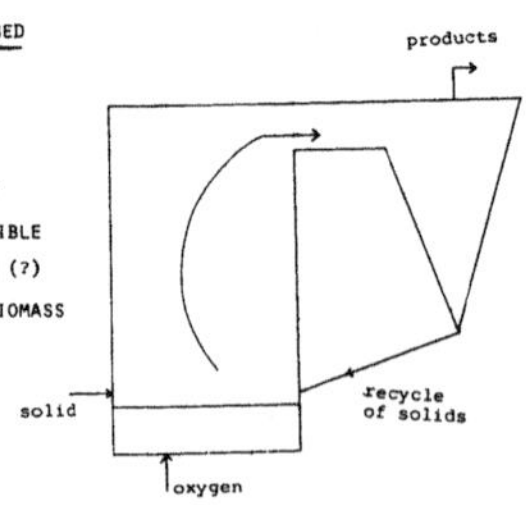

POWDER FLAME GASIFIER

+ VERY LOW TAR AND METHANE
+ FLEXIBLE TO FEEDSTOCK TYPE
- EXTREME SIZE REDUCTION AND FEED PREPARATION
- COMPLEX TO CONTROL

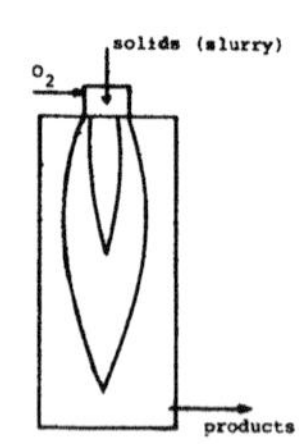

Figure 7. Overview of proposed reactor types; some advantages and disadvantages.

Cross current moving bed.

This reactor type avoids the permeability/channeling problems while it retains the favourable operation characteristics of a packed bed. The contacting pattern between gas and solids is not favourable from a point of view of pyrolysis products conversion but the reactor is used in a pyrolysis gas recycle mode in the proposal utilizing this system (Cemagref [15]). This may, increase pyrolysis product conversion.
In one proposal the cross flow reactor is used for pyrolysis of the wood (Foster Wheeler [17]). The tar and char resulting from it are mixed up in a slurry which serves as a feed for a powder flame gasifier.
This is an expensive method of feeding but, at least to a certain extent, this will be counterbalanced by the more favourable gas composition of product gas from this type of gasifiers

Fluidized bed reactor systems

A fluidized bed gasifier has been originally developed by Winkler for coal gasification. Such reactors can handle a wide range of feedstock including those with a difficult shape or poor mechanical strength of the char. Problems may be the limited pyrolysis product conversion, limited char conversion and the more complex solid flowrate control required.
Reactors utilizing multiple fluid beds rely on the conveying of solids from one vessel to another. Such systems are well proven for catalytical cracking. In gasification, however, flow properties of the bed material are much worse than those of specially designed cracking catalyst. Therefore this point still asks for attention.

Fast or recycling fluidized bed.

A fluid bed at extreme high gas velocities with continuous recycle of particles. It is claimed to have a higher flexibility and less ash melting problems than a normal fluid bed (important when gasifying with oxygen/steam). This would allow for higher temperatures (favourable for optimal product gas composition). The reactor originally was developed by Lurgi for coal gasification and is now proposed for wood gasification as well.

Powder flame gasifier.

Already known from oil and coal gasification (Koppers/Krupp; Shell; Texaco), this type of reactor is now also proposed for wood gasification (Foster Wheeler [17]). It's an interesting process because of feedstock flexibility and a very high operation temperature which may lead to product gas that is virtually free of hydrocarbons. Disadvantages are that this gasifier requires a complex and reliable solids feeding system and is possibly expensive and difficult to operate at relatively low capacities.

Cyclonic reactor

Specially designed for small particles and short contact times, it may provide excellent contact between gas and solids. This reactor is under development by Cemagref [21] but they have not yet made a proposal for scaling-up. On the long term however, development of such a system looks prospective. Figure 8 shows the principle, together with some advantages and present problem areas.

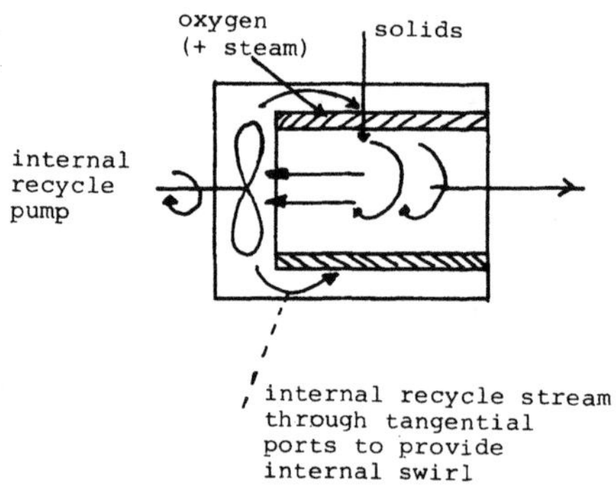

Figure 8. Cyclonic gasifier.

+ GOOD MASS TRANSFER
- SOLIDS ENTRAINMENT
- DIFFICULT TO SCALE-UP

The countercurrent reactor.

This reactor has not been proposed; probably not because of channeling and flow problems in the higher area's of the packed bed, especially with irregularly shaped biomass, producing much tar when operating in counter-current mode. Furthermore the large amount of tar produced requires extensive gas conditioning and complicates the operation. For well graded coal this reactor type has found widespread application.

The molten salt bath and the rotary kiln reactor.

These reactors have not been proposed, probably not because they are considered as being too complex and too expensive for biomass gasification.

Pressurized gasification.

As discussed above, pressurizing the gasifier reduces overall compression costs for methanol sysnthesis and therefore some pressurization is adventageous, with the optimum probably lying somewhere between 10 and 30 bar, depending on the type of process.
Four proposers (Lurgi, CJB-Wellman, ULB and Fritz Werner) prefer to develop their process for atmospheric pressure.
In all other proposals gasification takes place under pressure (8-30 bar). We wonder however wether in all cases the aims will

be realized within the C.E.C.'s tight time schedule (before 1984) and/or within the budget allocated.

Capacity of pilot plant.

All proposals comply with the condition of the C.E.C. of a minimum gasification capacity of 10 tons of wood/day. Some are even an order of magnitude larger.

Conclusions.

The proposals submitted to the C.E.C. cover an interesting variety of processes. Many of these proposals clearly aim not only at methanol synthesis but also contribute to further development of efficient biomass gasification systems for small and intermediate scale shaft power and electricity generation. This reduces the financial risks of a project because the economical feasibility will become less sensitive to an uncertain future methanol market. All projects aim at producing medium Joule gas from wood at a capacity of at least 10 tons of wood/day. Proposed methods include oxygen gasification, steam gasification and gasification with chemically bound oxygen produced in a separate reactor from air and a reacting solid. Moving bed gasifiers (both co- and cross current), fluidized bed gasifiers (both single- and multiple beds) a "fast" or circulating fluidized bed and a powder flame gasifier are proposed

References

[1] G. Ader, A.V. Bridgwater, B.W. Hatt in: W. Palz, P. Chartier and D.O. Hall (Eds.) Energy from Biomass, Appl. Sci. Publ., London (1981), 598.

[2] R. Channing Johnson, R.K. Lay and L.C. Newman, Energy from Biomass, A Technology Assessment of Terrestrial Biomass Systems, MITRE, Mc.Lean, Va, (1980).

[3] E. Nitscke and J. Keller, Chemie Technik, 9 (1980) No 3, 121.

[4] SERI ITR-33-239, A Survey of Biomass Gasification, Volume III, Current Technology and Research, SERI, Colorado, 1980.

[5] W.P.M. van Swaaij in: W. Palz, P. Chartier and D.O. Hall (Eds.), Energy from Biomass, Appl. Sci. Publ., London (1981), 485.

[6] T.B. Reed, A Survey of Biomass Gasification, Vol. I, SERI ITR-33-239-VI, Colorado (1979).

[7] A.E. Hokanson e.o., Chemicals from Wood, NTIS, PB-262489 (1975).

[8] Science Applications, Inc., Evaluation of the State of the Art of Biomass Based Methanol Processes. Mc.Lean, Va. (1978).

[9] R. Channing Johnson, Energy from Biomass, The Implications of Methanol from Silvicultural Farms, MITRE, Mc.Lean, Va. (1980).

[10] Kirk-Othmer, Encyclopedia of Chemical Technology, Wiley (1981) 15, 398.

[11] Centre de Recherches Industrielles de L'Université Libre de Bruxelles, ULB/FUL/CWB/FRW/ACSA, these proceedings

[12] John Brown-Wellman Mechanical Engineering Ltd., these proc.

[13] Creusot Loire, these proceedings.

[14] International Research and Development Co; Pollution Prevention Control, these proceesings.

[15] Creusot-Loire Entreprises/Cemagref, these proceedings.

[16] Novelerg, these proceedings.

[17] Foster Wheeler Products Ltd., these proceedings.

[18] Fritz Werner, Maremma, these proceedings.

[19] Lurgi Mohle und Mineralöltechnik GmbH, these proceedings.

[20] H.F. Feldmann, in Energy from Biomass and Wastes, IGT, Washington (1978).

[21] J.F. Molle in: W. Palz, P. Chartier and D.O. Hall (eds.) Energy from Biomass, Appl. Sci. Pub., London (1981), 574.

PART I - METHANOL FROM WOOD

Synthetic fuel from biomass : the AVSA dual fluid bed combustor - gasifier project

Gazéification du bois en lit fluidisé à l'oxygène et sous pression en vue de produire un gaz utilisable pour la synthèse du méthanol

Development of the oxygen donor gasifier for conversion of wood to synthesis gas for eventual production of methanol

Synthetic fuel from wood using steam and air

Experimental work on a fixed-bed oxygen gasifier in the view of methanol synthesis using biomass as a feedstock

Design and construction of a pressurized wood gasifier with heat input by oxygen combustion or by electrical heating

Proposed 20 tonnes per day biomass gasification pilot plant

Synthesis gas obtained from biomass

Gasification of wood in the circulating fluidized bed - methanol production route

SYNTHETIC FUEL FROM BIOMASS : THE AVSA DUAL FLUID BED COMBUSTOR - GASIFIER PROJECT.

Authors	: A. BARY, H.A. MASSON, P. DEBAUD.
Contract number	: ESE/P/001/B
Duration	: 24 months : 1 january 82 - 31 december 1984.
Total budget	: 50 10^6FB EEC contribution : 25 10^6FB
Head of project	: P. DEBAUD, President of AVSA.
Contractor	: Association pour la valorisation de la vallée de la Sûre et de l'Attert (AVSA).
Address	: AVSA c/o Roger KAUFFMANN 43 Les Frênes 6633 FAUVILLERS BELGIQUE

Summary

The AVSA project covers completely the generation of synthesis gas from wood waste : feed collection, sizing, drying and transportation as well as gasifier design.
As biomass is a dispersed material, the transportation costs increase sharply with the capacity of the plant. As a consequence, the reduction of methanol production cost with increasing size of the plant is not evident when biomass is used as feedstock.
For plant sizes ranging between 20 and 1000 T/D, we plan to define the optimal harvesting technique, the associated cost of wood waste and the forest surface needed in an european context.
The second aspect of our proposition is the design of a dual fluid bed gasifier-combustor. This system is characterized by the absence of 0_2, operation at atmospheric pressure and a great compacity. The circulation principle and tentative heat and mass balances ar presented.
Our options are discussed in connection with methanol synthesis. Some further developments need to be done parallely on a micropilote and on a real scale cold model.
The synchronization of these different aspects is finally presented as a time table.

1. INTRODUCTION

The present project is the AVSA (Association pour la valorisation de la Sûre et de l'Attert) response to an EEC Call for Tenders in the field of methanol fabrication from wood waste. It has been clearly stated that the goal of the work is the development of new gasifiers more than the realization of state of the art demonstration units.
The final result has to be the realisation of a new, competitive and reliable process. The project has to cover an experimental program which may include forestry engineering and feedstock preparation.
It is the merit of AVSA, located at Martelange in the south of Belgium, to have, firstly, understood the representativity of the Belgian forest as a sample of the EEC forests (table I).
Secondly, with the support of regional economico political personalities and organisations, to have proposed forestry valorisation as a solution to unemployement in the very south of Belgium.
Thirdly to have formed a consortium including mainly "Le Centre wallon du bois", which is the leader belgian center for forestry technology, Cockerill, a major belgian metallic constructor, and the center for industrial research of the Free University of Brussels where fluidization and fluidized bed gasification research is running for years now.
As a result of this association, we propose an integrated project covering all the aspects of the preparation of a synthesis gas able to feed a methanol plant. Forestry aspects in connection with cost and availability of the feedstock are investigated as well as the design and the operation of a 20 T/D fluid bed gasifier.

2. DESCRIPTION OF THE GASIFIER.

Gasification is a very endothermic process and calories must be injected in the system.
Classicaly, a part of the organic charge is burned with air and the heat produced by combustion is used.
Acting this way, one produces a gas which is largely diluted by N_2 and is unsuitable for synthesis.
The classical solution consists in using pure oxygen as combustion gas. This choice conducts to the erection of an expansive O_2 plant; it also leads to an operational temperature for gasification of about 1.200° C. This is a disadvantage when looking at thermal losses or at the choice of inexpansive construction materials.
In this project, combustion is achieved in a separate fluid bed operating at 900° C, using air as gas. A solid circulation loop is realized between the combustor and the gasifier. This last unit is also a fluid bed fed with steam and eventually with recycled purge gas. It works at about 700°C.
The transfered solid, mainly inert sand, ash and few percents of wood waste pieces (size reduced to less than 2 cm), acts as the heat source for the gasifier.
The originality of our process is in a simplification of the solid transfer lines leading to a more compact design.

The faisability of the circulation loop has been demonstrated at a 20 cm scale, on a cold model. Wood gasification and combustion data have been collected in a batch fluid bed (15 cm in diameter) operating between 650 and 900° C, and also in a thermobalance.
As in any fluidized bed system, demixing of the wood pieces could occur, with catastrophic consequences on the gas quality and yield. At a 20 cm scale, the correct operating conditions have been defined and gas leakage through the transfer lines is very small.
However, for safety we suggest, parallely to the design of the real scale gasifier, to make circulation and demixing tests on a real scale cold model of the system, using the different possible feedstocks.
At his scale, gas leakages shall also be investigated.
For both the demixing and leakage measurements, fully automatic experimental facilities exist at the University of Brussels and are directly connectable to the large scale cold model.
Construction of this model and realisation of the main tests are planned to be completed in four months.

3. PERFORMANCES OF THE SYSTEM.

Fig. 1 summarizes the heat and mass balances estimations for the gasifier. They were deduced from small scale tests, extrapolated with a lot of pessismim. The real figures could probably be more attractive.
From this figure, it is possible to deduce :

- the mass efficiency : 8,56 mole (H_2 + CO)/Kg wood (20 % moisture)
- the net inputs in the gasifier :
  - oxygen : 0.0 kg/kg wood (20 % moisture)
  - steam (400° C, 1.2 atm.) : 0.3 kg/kg wood (20 % moisture)

The specific electricity consumption is expected to be no more than 0,100 Kwh/kg wood (20 % moisture).
An indicative flow sheet of the system is given in fig. 2.
Gas purification is done by cyclones and water scrubbing to remove tar.
Other noticable points are :

- use of the combustion flue gas for :
  - preheating of combustion air;
  - preheating of water supply for steam gasification;
  - drying of the raw feedstock in a fluid bed dryer.
- heat recovery from the discharged solid to warm up recycled scrubbing condensates.
- use of the methanol synthesis purge gas as complementary source of heat in the combustor.

Complementary heat shall be electrically supplied in the demonstration unit, for simplicity.

4. COMPATIBILITY OF THE GASIFIER WITH A METHANOL SYNTHESIS PLANT.

a) Operation at atmospheric pressure.

Table II recalls the advantages and disadvantages we see in

operating pressurized gasifiers. As we dit not find decisive advantages in an increase in pressure, we choiced to work at atmospheric pressure. This way, the design of the gasifier is simplified and all the extrapolation risks are only focused on size increase.

b) CH_4 content of the synthesis gas.

Table III synthetizes our views on this problem. It is important to make a distinction between CH_4 generation in the methanol reactor and CH_4 content in the synthesis gas.
The first event is critical for the catalyst live. The second acts mainly on the compression costs. Indeed CH_4 behaves as an inert in the synthesis of methanol.
But, a high CH_4 content allows to use the purge gas as heat supply by recycling it to the combustor-gasifier system.
It asks for selecting a methanol synthesis process having an as high as possible conversion per pass, to minimize the (CO / H_2) losses by purging.

c) Tar content.

The amount of tar formed is an important factor because it is related to the yield of gas generated and to compression difficulties of the synthesis gas.
It is mainly controlled by the gasification temperature.
As the combustion temperature must remain under 900° C, the temperature of the gasifiers ranges between 700 and 750° C, where 5 to 10 % of tar is expected to be formed.
Tar shall be removed by cooling and water scrubbing of the gas.
The condensed mixture is reinjected as gasification agent : it is warmed up by the purge gas, by the steam generated in the methanol synthesis reactor and by heat recovery from the discharged solid.

5. INDICATIVE COST FIGURES.

A preliminary economic evaluation has been made for a 20 T/D and a 1.000 T/D unit (table IV).
At his stage of development, this information must be considered as defining an order of magnitude only.
However, it appears clearly that the use of biomass for generating synthesis gas is an expansive solution.
To reduce the cost of the produced gas, one has to act on the cost factors. The most accessible are the direct costs.
Utilities and feedstock cost become dominant factors when the plant capacity increases.
A major part of the detailled design of the gasifier shall concern the minimization of the needed utilities.
At this stage also, we have estimated the feedstock price to be of about 1.200 FB/T for both the plant sizes. This value has a confidence interval of about 50 % and would increase with the plant capacity. Indeed the transportation costs increase rapidly with the wood waste collection radius needed.
To precise the wood waste costs as well as the optimal gasifier size, we plan to realize a large harvesting research and development work.

6. FEEDSTOCK EVALUATION.

Biomass waste may be formed as by product of forest engineering and also in wood transformation industries.
In the EEC forests, one may collect coppies yet present on large surfaces and also parts of the production of high forest (thining young trees, top branches or lodging slash) completely neglected until now.
In wood transformation industries, saw dust and bark are the less used products, and are thus potential feedstocks for new processes. Other kinds of waste are chipped and are mainly transformed in agglomerated pannels or paper pulp.
Industrial waste is a more homogeneous and localized material than forestry waste, and is thus more appreciated as feedstock.
However large quantities of woody mass are available in the forest.
But this material must be economically collected, sized and dryed before becoming a valuable gasifier feedstock. Actually very few technical and economical analyses of this problem have been done.
It appears to us, in this context, that it is of prime importance to develop appropriate harvesting, sizing and drying techniques, for several feedstocks of the gasifier (hardwood, softwood, stem, branches, chips, saw dust, ...).
As said in the introduction, the south of Belgium is a representative sample of EEC forests. A preliminary study indicates that it is possible to collect, in this region, 100.000 tons per year of dry material, in a radius of 75 km. This corresponds to a 330 ton per day plant.
Fig. 3 is a synthesis of our proposal in this field.
All the three aspects of harvesting would be studied (felling, skidding and chipping) as well a sizing and drying.
To be significative, this study must also cover for the several materials, screw feeding and gasification tests. For this we plan to send those materials in a small scale (30 cm in diameter) unit and also to study their circulation and demixing characteristics in the large scale cold model yet described.
At the end of this study, the best feed material shall be selected as well as the associated harvesting, sizing and drying techniques.
These results shall be used to feed the 20 T/D plant during the 15 days run and also during preliminary tests of the gasification unit.
The needed weights of solid are also given in fig. 3 for the different steps of this program.

7. TIME TABLE.

Table V shows the general lay out of the project organisation.
The small scale tests and most of the model runs may be done independantly and in parallel with the design, the construction and operation of the 20 T/D plant. Harvesting techniques are developed parallely and interact partially with small scale and cold model operations.

8. DEFINITIVE FINANCIAL ARANGEMENTS.

Table VI summarizes the stituation existing the 21th of october 1981.
Any further development shall be immediatly communicated to the EEC.

TABLE I

REPRESENTATIVITY OF THE SOUTH-BELGIAN FOREST

		South Belgium	EEC average
Foresty coverage (%)		30	20
Forest surface/habitant (ares/m)		12	12
Total productivity (m^3/ha/year)		5	3
Species	softwood (%)	44	40
	hardwood (%)	56	60

TABLE II

EFFECT OF PRESSURE ON THE COMPATIBILITY WITH METHANOL SYNTHESIS

Advantages	Desadvantages of pressure
1. Higher gasification yield (factor 1.5 from 1 to 20 atm.) 2. Reduction of compression costs. 3. Increase in throughput per unit section.	1. More expensive gasifier. 2. Feeding problems. 3. Combined size and pressure extrapolation problems. 4. Greater CH_4 production. 5. Higher elutriation rate. 6. Lower gasification rate (factor 3 between 1 and 20 atm.)

TABLE III

PROBLEMS ASSOCIATED WITH CH_4 CONTENT.

Main problem : avoiding CH_4 formation in the methanol reactor (very exothermic : Ni catalyst sintering).

If CH_4 is yet present in the gas :

1. decreases the mass conversion to synthesis gas;
2. increases the compression costs;
3. increases the synthesis gas lossed by purge.

However :

1. CH_4 formation limitated in the methanol reactor (equilibrium).
2. Purged gas is usefull (high content of CH_4) = auxiliary fuel for the combustor.
3. Select a methanol synthesis process with a conversion per pass as high as possible (i.e. chemsystem 3 phase fluid bed reactor) : (CO, H_2) content of the purge is low.

TABLE IV

ECONOMICAL DATA FOR SYNTHESIS GAS (ORDER OF MAGNITUDE)

Capacity (T/D)	20	1.000
Mass efficiency (%)	50	50
Synthesis gas production (T/D)	10	500
Total capital cost (10^6 FB)	68	1.080
Total annual operating costs (10^6 FB)	46	1.136
Estimated gas price		
FB/kg	13,71	6,74
FB/KJ	0,86	0,42
Structure of direct costs (%)		
labor	28,6	0,8
feedstock (1.200 FB/T)	28	42,3
utilities	37,4	56
others	6	0,9

TABLE V - TIME TABLE

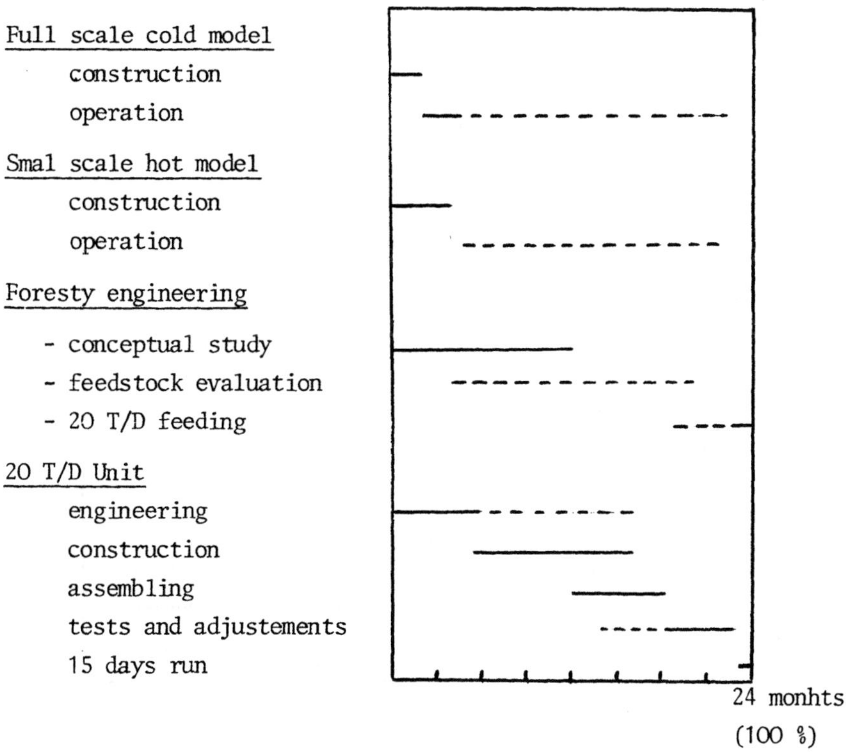

TABLE VI

DEFINITIVE FINANCIAL ARRANGEMENTS (10^6 FB) ENGAGED BY THE CONSORTIUM

Total cost :	50		
Possible EEC contribution	-25		
	25		
Centre wallon du bois		4.3	
University of Brussels		5	
Cockerill		5	
Cellule de gestion des contrats technologiques		0 - 12.5	(*)
	25	14.3 - 26.8	

(*) following potential exportation market for the gasifier. The C.G.C.T. covers 50 % of the market analysis study.

Fig.1. MASS BALANCE. (kg/hr).

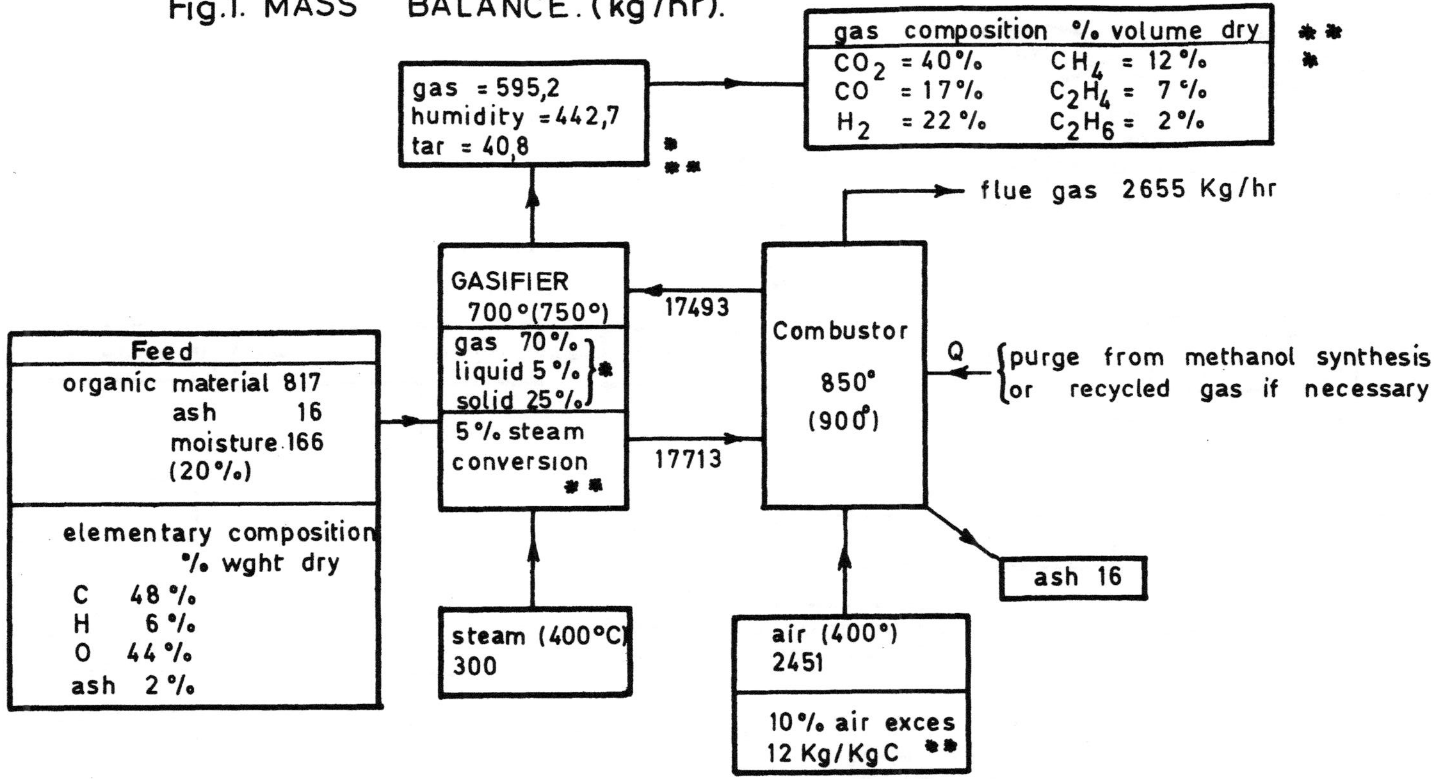

* adapted from PYROX system (KUN II) using urban waste

** measurements done in a 15 cm diameter bed (VUB) working as combustor or steam gasifier

Fig.2 FLOW SHEET.

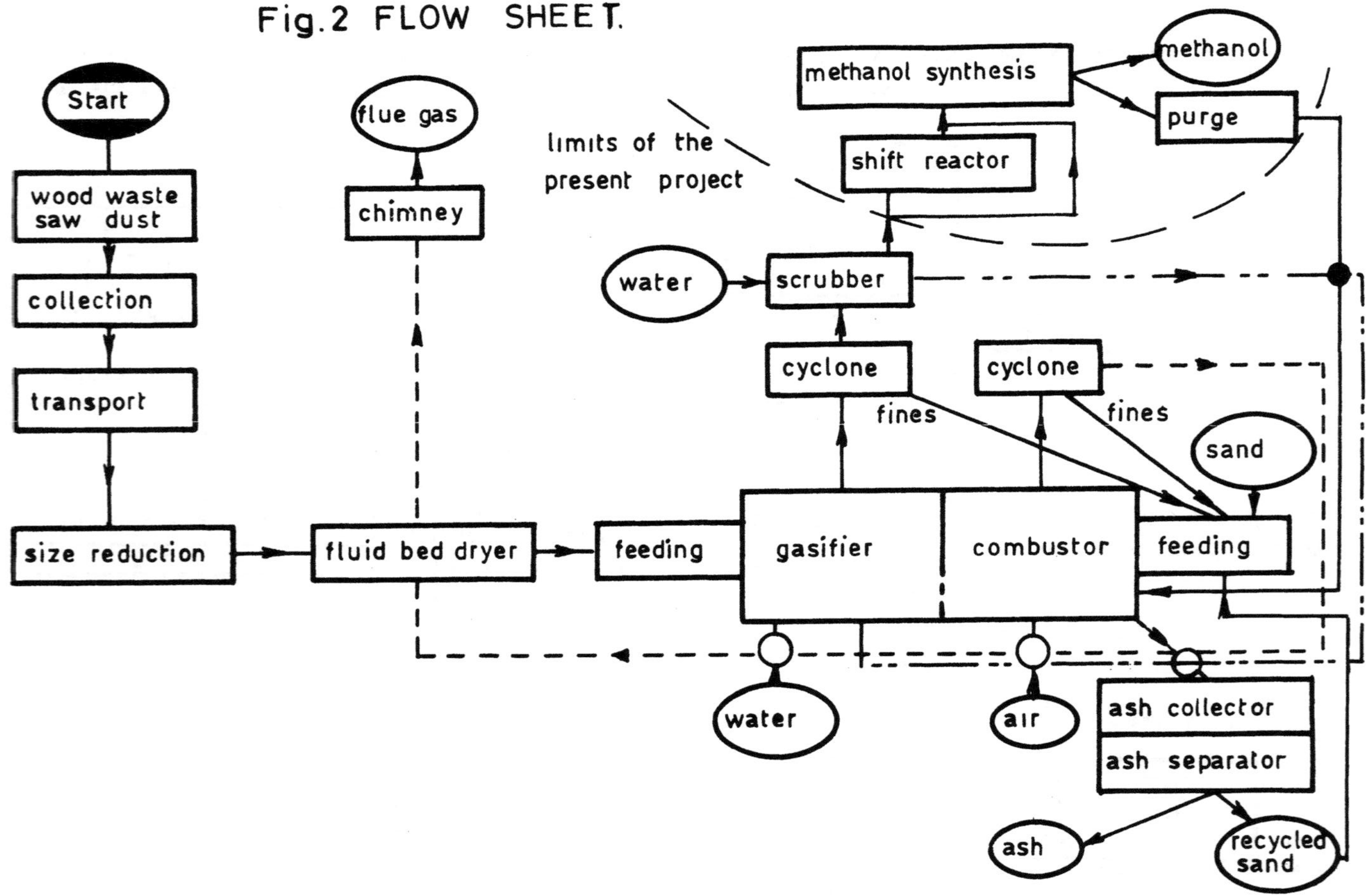

Fig. 3. FEEDSTOCK SOURCES.

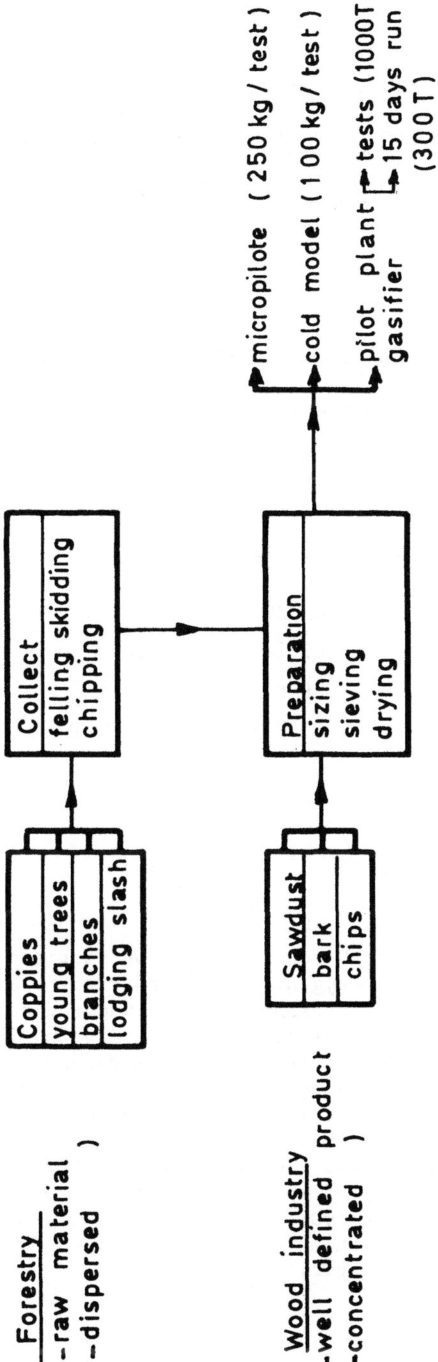

GAZEIFICATION DU BOIS EN LIT FLUIDISE A L'OXYGENE ET SOUS PRESSION EN VUE DE PRODUIRE UN GAZ UTILISABLE POUR LA SYNTHESE DU METHANOL

AUTEURS : G. CHRYSOSTOME - J.M. LEMASLE

N° DE CONTRAT : ESE/P/004/F

DUREE : 15 mois

CHEF DE PROJET : G. CHRYSOSTOME

CONTRACTANT : CREUSOT-LOIRE

ADRESSE : CREUSOT-LOIRE
Division Energie
BP N° 31
F 71208 LE CREUSOT

SUMMARY

The present research scheme covers pressurized fluidized bed wood gasification with oxygen. It is supported by a research experiment being conducted at present with a fluidized bed reactor allowing to gasify 100-150 kg/h of dry wood with oxygen. The intended reactor will make it possible to gasify 60 tons/day of dry wood under a 15 bar pressure. The equipment is dimensioned so that it can operate under pressures of between 10 bar and 30 bar. The demonstration plant includes all the facilities permitting to operate it under industrial conditions : biomass drying; feeding biomass into the gasifier with help of an uninterruptedly operating machine; heat recovery ; dust removal ; recycling of unburnt carbon if any. Different kinds of biomass can be tested : wood, wastes from farming or from forestry activities.

1 - INTRODUCTION

Actuellement, la synthèse du méthanol est réalisée industriellement à partir du gaz naturel ou de fractions pétrolières. Les unités industrielles qui peuvent produire jusqu'à 1500 t/j de méthanol, comprennent en général :

- la fabrication d'un gaz de synthèse brut ($CO + H_2$)
- l'épuration du gaz de synthèse et l'ajustement de sa composition ($H_2/CO=2$)
- la synthèse proprement dite qui s'effectue sous pression moyenne (50 à 100 bar).

La gazéification de la biomasse qui permet l'obtention d'un mélange de monoxyde de carbone et d'hydrogène devrait dans un proche avenir conduire à des productions importantes de méthanol.

La synthèse du méthanol est effectuée industriellement avec un gaz aussi pur que possible, il est donc indispensable de gazéifier la biomasse à l'oxygène pur. Compte-tenu de ce qui précède, la synthèse doit s'effectuer sous pression.

CREUSOT-LOIRE exploite actuellement dans son Laboratoire d'Essais Energétiques, un réacteur permettant la gazéification en lit fluidisé et à l'oxygène d'environ 150 kg/h de bois sec. Nous proposons maintenant une extrapolation raisonnable de ce gazogène (environ 10 à 15 fois) qui conduira à une usine pilote permettant la gazéification, à son débit nominal, de 60 t/j de bois sec sous une pression de 15 bar.

Le gazogène à lit fluidisé développé par CREUSOT-LOIRE fonctionne actuellement avec des copeaux de bois (plaquettes papetières), cependant des essais sont prévus avec des sciures, des déchets d'exploitations forestières et agricoles. Afin d'introduire ces différents produits dans le gazogène nous développons actuellement des machines spéciales permettant l'injection de la biomasse dans des réacteurs sous pression. Il est prévu que la même machine pourra fonctionner aussi bien avec des copeaux de bois, qu'avec de la sciure, de la paille ou du bois déchiqueté.

2 - DONNEES DE BASE

La présente proposition de Recherche s'appuie sur une étude expérimentale effectuée par la Société CREUSOT-LOIRE dans son Laboratoire d'Essais Energétiques du CREUSOT. Le gazogène à lit fluidisé, objet de l'étude, bénéficie de subventions des Communautés Européennes et du Commissariat à l'Energie Solaire. Cet appareil a déjà permis la gazéification à l'oxygène et sous pression atmosphérique de 10 tonnes de bois alimentées sous forme de plaquettes papetières. Des essais de gazéification de paille à l'oxygène sont actuellement en cours et des résultats significatifs portant sur 2 tonnes de matériaux, seront disponibles avant fin octobre 1981. Dans sa version sous pression atmosphérique, l'appareil permet la gazéification à l'oxygène pur d'environ 150 kg/h de bois sec. La description de l'installation de gazéification est présentée dans l'annexe 1 ci-après. Le dimensionnement du gazogène, objet du présent avant-projet de recherche et décrit aux paragraphes suivants a été déduit des résultats obtenus au moyen de l'appareil expérimental du Creusot.

2.1 - RESULTATS EXPERIMENTAUX

Environ 20 essais ont été effectués à ce jour sur des plaquettes de bois, à l'oxygène pur. Les caractéristiques de ces essais sont résumées ci-dessous.

Conditions opératoires :

. Bois :	plaquettes papetières	= 3 à 40 mm
	épaisseur max	= 10 mm
	essence	= pin
	humidité	= 20 à 40 %
	capacité	= 90 à 110 kg/h (bois sec)
. Oxygène :	pureté	= 99,5 %
	débit	= 18 à 22 Nm^3/h
. Lit fluidisé :	sable extrasiliceux	= 98,5 % SiO_2 min.
	granulométrie	= 0,125 à 0,600 mm
	charge	= 50 kgs

. Température :   dans le lit fluidisé = 800 °C
en sortie réacteur = 700 °C

Résultats :

Chaque essai a une durée, en régime stationnaire comprise entre 1h30mn et 3h30mn. La synthèse des résultats expérimentaux nous conduit à la composition moyenne suivante pour le gaz sec.

CO = 35,5 % volume
CO^2 = 30,3 %
H_2 = 18,7 %
CH_4 = 10,0 %
N_2 = 1,8 %

Débit gaz humide = 150 Nm^3/h (pour un débit de 130 kg/h à 20 % d'humidité)

Débit gaz sec = 102 Nm^3/h

Taux de conversion du carbone = 85 %

3 - CHOIX DES CARACTERISTIQUES DU REACTEUR DE GAZEIFICATION

3.1 GAZEIFICATION A L'OXYGENE PUR

Le mélange monoxyde de carbone, hydrogène envoyé à la synthèse du méthanol doit être le plus pur possible ; il faut donc éviter la présence d'azote et gazéifier à l'oxygène pur ou à la vapeur d'eau. Cependant, la gazéification à la vapeur d'eau suppose un apport de chaleur par les parois du réacteur (chauffage électrique par exemple) ou par circulation d'un solide caloporteur inerte chimiquement ou non; ce qui suppose alors un second réacteur où est réalisé l'apport de chaleur au solide. La solution retenue a été la gazéification à l'oxygène pur.

3.2 PRESSION

Toutes les études effectuées montrent qu'il y a intérêt, d'un point de vue énergétique, à produire le gaz de synthèse sous la pression la plus élevée possible. Les générateurs actuels de gaz utilisant des coupes pétrolières ou du gaz naturel travaillent sous une pression de 30 bar.

Par ailleurs, les études évoquées précédemment montrent que le gain énergétique est déjà important à partir de 10 bar. Compte-tenu des incertitudes demeurant quant à la possibilité d'introduire de la biomasse dans un réacteur maintenu à 30 bar, une plage de fonctionnement de 10 à 30 bar a été retenue pour le gazogène.

3.3 - DEBIT DE BOIS GAZEIFIE

Les lits fluidisés ont cette caractéristique importante que la vitesse des gaz doit être pratiquement constante quelle que soit la pression du réacteur. Par exemple, sous 20 bar, le même gazogène à lit fluidisé permettra de gazéifier environ deux fois plus de bois qu'à 10 bar. Compte-tenu de cette caractéristique, le réacteur a été calculé de telle façon qu'il permette la gazéification de 2,5 t/h de bois sec (60 t/j) sous 15 bar absolus. Le débit est environ 10 fois plus important que celui du réacteur actuellement en fonctionnement et nous paraît représenter une extrapolation raisonnable. Sous 10 bar, limite inférieure de l'intervalle de pression, le débit de l'appareil ne serait plus que de 40 t/j de bois sec. Sous 30 bar, limite supérieure de l'intervalle de pression, le débit de bois sec pourrait atteindre 5 t/h (120 t/j).

4 - DESCRIPTION DE L'INSTALLATION PILOTE

L'installation pilote est décomposée en 10 sections. La figure 1 en représente un schéma simplifié des principales étapes.

Section 1 - Réception et stockage du bois (ou plus généralement de la biomasse). Il est admis que le gazogène pilote sera installé dans une usine utilisant déjà du bois, notamment papeterie, et que les installations existantes pourront être utilisées.

Section 2 - Production de l'oxygène. Il est envisagé soit d'utiliser de l'oxygène liquide livré en containers, soit de produire de l'oxygène sur le site. Les conditions locales détermineront laquelle des deux solutions sera retenue.

CREUSOT-LOIRE

USINE DU CREUSOT

DIVISION ENERGIE

PRESSURIZED OXYGEN BLOWN FLUIDIZED BED GASIFIER

PROJECT CONTENTS

Capacity = 60 TPD Dry Wood

Pressure = 15 Atm

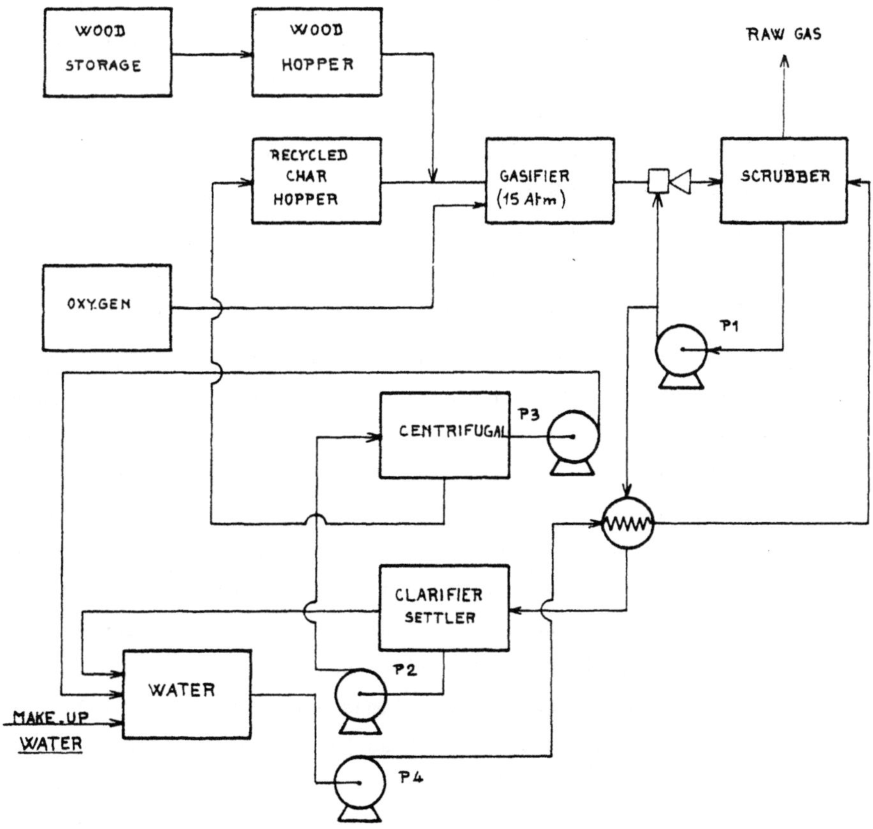

FIGURE 1

Section 3 - Production des utilités. Il est admis que la vapeur éventuellement nécessaire pour le séchage du bois et le fonctionnement du gazogène, l'eau, l'énergie électrique et de façon générale les utilités nécessaires, seront fournies par le site.

Section 4 - Préparation de la biomasse. Elle comprend un broyage du bois et une élimination des corps étrangers (pierres, ferrailles) par séparation magnétique et séparation pneumatique. Le bois convenablement calibré est ensuite séché (humidité ramenée à 15 % si nécessaire) puis dirigé au moyen d'un transporteur vers une trémie de stockage.

Section 5 - Machine d'introduction de la biomasse dans le gazogène.

La biomasse prélevée à la base de la trémie de stockage est introduite directement dans le gazogène au moyen d'une machine spéciale. Cet appareil fait actuellement l'objet de tests dans les laboratoires de CREUSOT-LOIRE ; il permet d'introduire en continu de la biomasse dans un réacteur sous pression. Son fonctionnement ne se limite pas au bois sous forme de plaquettes papetières, il doit permettre en particulier l'introduction de paille, de sciures et de résidus d'exploitation forestière dans un réacteur sous pression.

Section 6 - Gazogène à biomasse fonctionnant à l'oxygène et sous pression. Il s'agit d'un réacteur à lit fluidisé qui se présente sous forme d'un cylindre vertical en acier, garni intérieurement de réfractaires. A sa base, le réacteur est muni d'une grille spéciale qui permet de distribuer uniformément l'oxygène pur et de supporter le lit fluidisé. Selon une conception propre à CREUSOT-LOIRE, le lit fluidisé est constitué d'une couche de matériau inerte à l'intérieur de laquelle est réalisée la gazéification de la biomasse. Cette disposition particulière permet de s'affranchir de la taille des morceaux de bois ou de biomasse et autorise l'utilisation de sciures, de déchets, de plaquettes papetières... De plus, cette disposition permet d'éviter la formation de goudrons. Le règlage du rapport oxygène-biomasse permet de contrôler la température du lit fluidisé. Un règlage fin de la température peut être obtenu grâce à l'injection d'une faible quantité de vapeur d'eau.

Section 7 - Conversion du méthane. Le gaz de synthèse brut quitte le gazogène à une température comprise entre 800 et 1000 °C. Le rechauffage de ce gaz à 1300 °C au moyen d'une injection d'oxygène permet de réduire la teneur en méthane à moins de 0,5 %.

Section 8 - Dépoussiérage du gaz de synthèse. Le gaz de synthèse contient les cendres du bois, il peut également contenir une quantité non négligeable de carbone. Le dépoussiérage sera effectué en fonction de l'utilisation du gaz de synthèse sur le site. Le mode de dépoussiérage envisagé est du type dépoussiérage humide par lavage à l'eau. Le dépoussiérage à sec imposerait en effet d'opérer à chaud; cette technologie n'est pas encore assez éprouvée sur le plan industriel pour ce type d'application.

Section 9 - Traitement des cendres - Recyclage du carbone imbrûlé. Les cendres et éventuellement le carbone sont récupérés dans le dépoussiéreur humide. Ils sont extraits au moyen de sas. Il est prévu de se réserver une possibilité de recycler le carbone imbrûlé si cette production était importante. Le recyclage du carbone vers le gazogène serait effectué au moyen de la machine spéciale d'introduction de la biomasse.

Section 10 - Epuration chimique et ajustement de la composition du gaz de synthèse. Cette section est indiquée pour mémoire. La recherche proposée ne comporte pas, pour l'instant, la synthèse du méthanol.

L'ensemble de l'installation pilote est automatisé et régulé de façon à :

. obtenir un fonctionnement stable, parfaitement contrôlé et en toute sécurité,

. être représentatif d'une installation industrielle.

5 - CARACTERISTIQUES PRINCIPALES DE L'INSTALLATION PILOTE

- Pression de fonctionnement : 15 bar.
- Consommation de bois : 60 t/j comptées sur sec.
- Composition du gaz de synthèse brut :

CO	29 %
CO_2	22 %
H_2	21 %
H_2O	20 %
CH_4	8 %

Nous avons étendu les caractéristiques de l'installation pilote projetée à une unité de production de méthanol complètement intégrée, afin de préciser le schéma de procédé et les bilans matière et énergie globaux.

Le schéma de procédé est représenté sur la figure 2 ci-après.

Dans cette configuration, les caractéristiques principales de l'unité seraient les suivantes (pour un débit de bois sec de 60 TPJ au gazéifieur)

. Consommation de bois = 65,82 t/j (dont 5,82 t/j pour alimenter la chaudière)

. Consommation d'oxygène = 39,1 t/j

. Usine entièrement autonome du point de vue énergétique.

. Production de gaz brut au gazogène = 94 730 Nm^3/j.

. Production de méthanol = 28,23 t/j

. Rendement thermique (PCI (méthanol)/PCI (bois gazéifié)) = 53,4 %

L'autonomie énergétique de l'unité est obtenue grâce à la récupération de chaleur à différents stades du procédés. L'appoint nécessaire pour "boucler" le bilan énergétique global de l'usine est réalisé en brûlant une partie du bois dans une chaudière générant de la vapeur ; cet appoint représente environ 10 % du débit de bois servant à la gazéification.

- Production potentielle de méthanol :

  1175 kg/h pour une usine intégrée

  1300 kg/h pour une usine recevant de l'électricité

- Rendement global de la synthèse du méthanol à partir de la biomasse production d'oxygène comprise : 54 %.

CREUSOT-LOIRE

USINE DU CREUSOT

DIVISION ENERGIE

PRESSURIZED OXYGEN BLOWN

FLUIDIZED BED GASIFIER

WOOD TO METHANOL CONFIGURATION

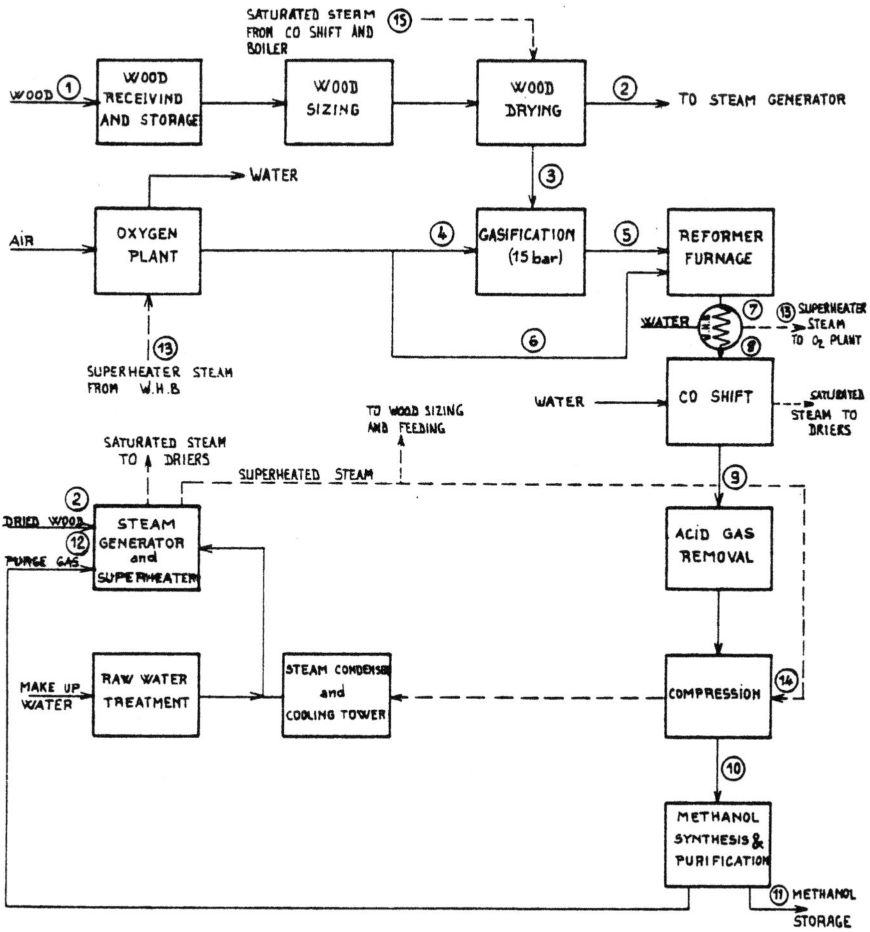

FIGURE 2

6 - PLAN D'EXECUTION

Durée totale : 15 mois

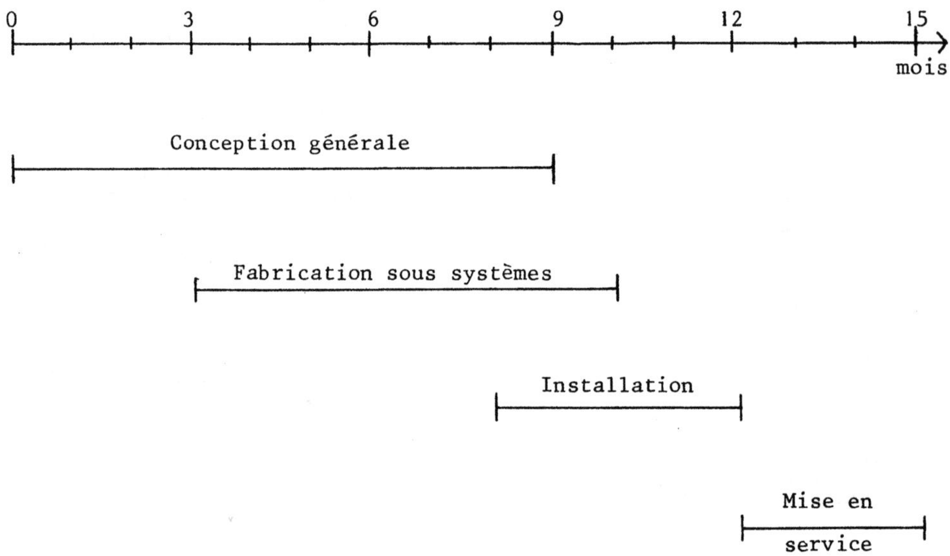

ANNEXE 1

DESCRIPTION DE L'INSTALLATION DE GAZEIFICATION PILOTE DU LABORATOIRE D'ESSAIS ENERGETIQUES DU CREUSOT

- Le schéma du gazogène est représenté sur la figure 3.
- Le plan de l'ensemble complet "gazogène + chambre de combustion" est représenté sur la figure 4.

L'appareillage comprend essentiellement un gazogène à lit fluidisé de diamètre intérieur 400 mm et une chambre de post-combustion des gaz produits au gazogène de diamètre intérieur 800 mm.

Afin d'éviter les longues mises en régime stationnaire, un appareil à faible inertie thermique a été choisi. Le gazéifieur est donc composé d'une enveloppe métallique (acier inox réfractaire) calorifugée extérieurement et muni à sa base d'une grille perforée permettant de supporter le lit fluidisé.

Le bois est alimenté dans une trémie de 0,7 m^3 avant d'être introduit dans le réacteur au moyen d'une vis sans fin à vitesse de rotation variable pour le règlage du débit.

Le matériau constituant le lit fluidisé est un sable extrasiliceux inerte, de granulométrie comprise entre 125 et 600 micromètres. La quantité introduite à chaque essai est de 50 à 60 kgs. La solution du lit fluidisé composé d'inertes permet la gazéification de morceaux de bois de dimensions très variées.

En début d'essai, la montée en température du réacteur s'effectue par la combustion de gaz naturel avec de l'air injectés dans le lit fluidisé. Un brûleur pilote disposé au dessus du lit assure la sécurité de cette opération. On porte ainsi le réacteur à 800 °C environ.

L'oxygène nécessaire à la gazéification est introduit à la base du lit fluidisé par une canne d'injection débouchant à 100 mm au dessus de la grille de fluidisation. C'est cet oxygène qui, lorsqu'il est injecté seul, assure la fluidisation.

Une canne d'injection d'eau est également disposée dans le lit fluidisé, permettant un débit de 0 à 100 l/h.

Le réacteur est muni, sur toute sa hauteur de 18 prises de température (thermocouples chromel-alumel) permettant de tracer la carte thermique de l'installation. Des prises de pression permettent de contrôler la pression au dessous et au dessus du lit fluidisé. Le maintien, au dessus du lit fluidisé, d'une pression légèrement inférieure à la pression atmosphérique permet d'éviter l'émission de gaz nocifs dans l'ambiance de la station.

Une fraction des gaz quittant le gazogène est dérivée vers les appareils d'analyse. La composition du gaz est mesurée au moyen d'un analyseur en continu à absorption dans l'infrarouge pour CO et CO_2; d'un analyseur en continu utilisant les propriétés paramagnétiques de l'oxygène pour O_2; d'un chromatographe en phase gazeuse pour CH_4, H_2, N_2 et Ar. Un traçage à l'argon est en effet réalisé lors des essais, permettant à partir d'un débit connu d'argon injecté de mesurer le débit de gaz produit.

La fraction restante du gaz (la plus importante) est brûlée au moyen d'un brûleur à gaz pauvre dans la chambre de combustion annexe, garnie de réfractaires. Les gaz de combustion sont ensuite dilués puis filtrés et enfin rejetés à l'atmosphère.

Le réacteur est également complété d'un cyclone.

Tous les débits de fluides sont mesurés au moyen de rotamètres ou de débitmètres à diaphragme. Le débit de bois est mesuré par pesée; son humidité est mesurée par prélèvement d'échantillons qui sont étuvés à 120 ° durant 12 h puis pesés.

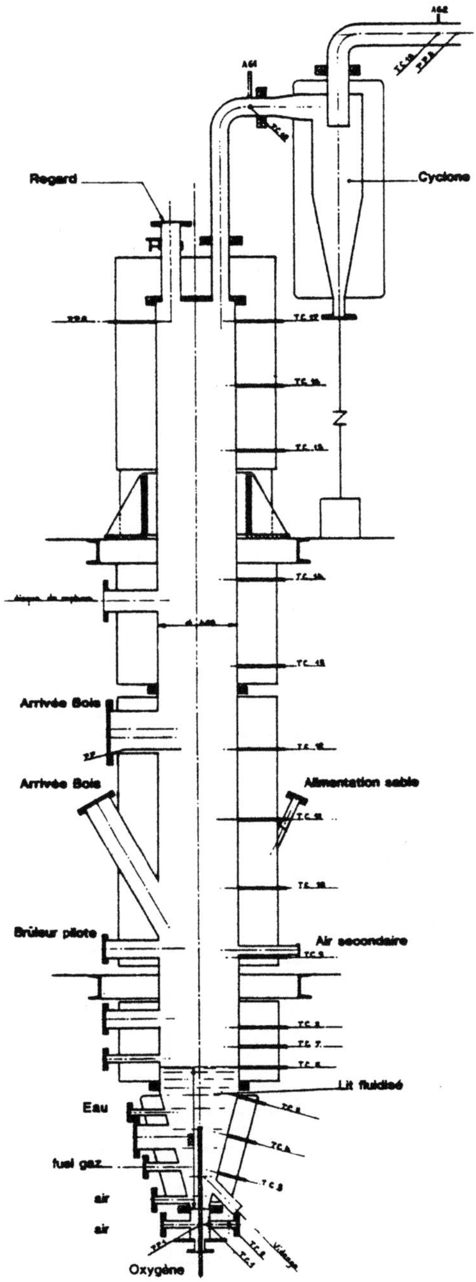

GAZEIFIEUR A BOIS
A L'OXYGENE
EN LIT FLUIDISE

Légende:
TC1 à 19 : Thermocouples
PP1 à 3 : Prises de pression
AG1 et 2 : Prises d'analyse gaz

FIGURE 3

PLAN DE L'INSTALLATION DE GAZEIFICATION DU BOIS A L'OXYGENE

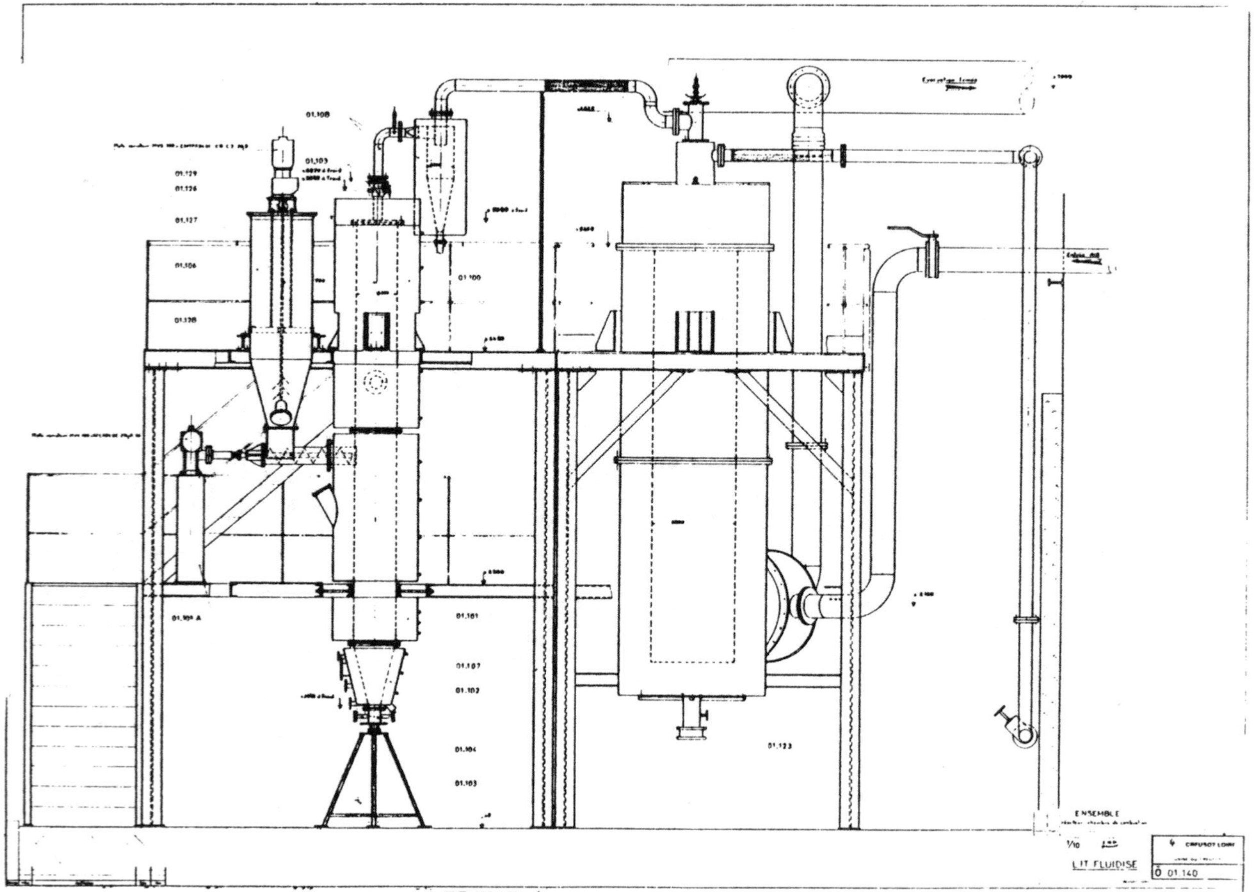

FIGURE 4

DEVELOPMENT OF THE OXYGEN DONOR GASIFIER FOR CONVERSION OF WOOD TO SYNTHESIS GAS FOR EVENTUAL PRODUCTION OF METHANOL

Authors : R.S. BICKLE, Dr. A.J. EDWARDS, Dr. G. MOSS

Contract Number : ESE/P/003/UK.

Duration : 24 months 1 Jan. 1982 - 31 Dec. 1983

Total budget : £ 621,000 CEC Contribution £ 300,000.

Head of Project : Mr E.J. BAVISTER, John Brown Engineers & Constructors Ltd

Contractors : John Brown Engineers & Constructors Ltd
Wellmann Mechanical Engineering Ltd

Address : John Brown Engineers & Constructors Ltd
Eastbourne Terrace, London W2 6LE

SUMMARY

The Oxygen Donor Gasifier consists of two reactors containing fluidised beds which are exchanged between the reactors in a continuous flow.

The wood is fed into the Gasifier which contains calcium sulphate and calcium oxide in its bed. The calcium sulphate is reduced to calcium sulphide providing the oxygen to gasify the wood. The gases leaving the Gasifier are cleaned in cyclones, cooled in a waste heat boiler, and finally cooled and cleaned in a direct water spray. Part of this gas is used to fluidise the Gasifier, the remainder being the product Synthesis Gas.

The stone from the Gasifier is transferred to the Oxidiser where the calcium sulphide is oxidised by air back to the sulphate. The hot flue gases leaving the Oxidiser are cleaned, and used to preheat the incoming air, the remainder of the heat in the flue gases can be used to predry the wood feed.

OXYGEN DONOR GASIFIER

Introduction

The Oxygen Donor Gasifier is a recent modification of the Chemically Active Fluidised Bed Gasifier (CAFB) which was developed on a pilot plant scale at the Esso Research Station, Abingdon (1,2,3) and demonstrated on a utility boiler in Texas by the Foster Wheeler Corporation. (4,5)

The C.A.F.B. Gasifier produces a desulphurised low BTU gas from liquid or solid fuels (6) and operates well on lignite. It comprises two reactors containing fluidised beds of lime, one a gasifying reactor and the other a regenerating reactor. Bed material is exchanged between these two reactors in a continuous flow, and both beds are fluidised with air.

In the gasifier the fuel is partially burnt at a temperature around 900°C to a low BTU gas, and the major part of the sulphur reacts with the lime to form calcium sulphide

$$2\ Ca\ O + 2S \rightleftharpoons 2\ Ca\ S + O_2$$

In the regenerator, the calcium sulphide is oxidised back to calcium oxide and sulphur dioxide

$$2\ Ca\ S + 3O_2 \rightleftharpoons 2\ Ca\ O + 2SO_2$$

The regenerator operates at a temperature around 1050°C, and the sulphur dioxide can be collected at quite high concentrations and in the Texas plant was subsequently reduced to elemental sulphur.

During this development work on how to maximise the yield of sulphur dioxide, it was found that with excess air and lower temperatures, very little or no sulphur dioxide was produced and the calcium sulphide was oxidised to calcium sulphate (7)

$$Ca\ S + 2O_2 \rightleftharpoons Ca\ SO_4$$

and that the oxygen fixed in this way and conveyed to the gasifier oxidised the fuel, reforming the calcium sulphide. This was demonstrated on the CAFB during periods of maloperation and is the basis for the oxygen donor gasifier, for which patent applications have been made by Exxon Research and Engineering Company. (8)

Initial feasibility tests with a fixed bed of calcium sulphate mixed with wood charcoal showed that, in the presence of hydrogen, calcium sulphate will oxidise wood charcoal to make a gas rich in carbon monoxide. The hydrogen had acted as a carrier for the oxygen from the calcium sulphate to the charcoal.

This batch work was followed by tests on the pilot plant scale during which residual fuel oil was gasified in a continuous fashion producing a nitrogen free fuel gas containing H_2, CO and CO_2.

Proposed Pilot Plant

The proposed Oxygen Donor Gasifier pilot plant for wood gasification is shown in Figure I and consists essentially of two fluid bed reactors, a Gasifier and an Oxidiser, arranged for interchange of the fluid bed material between the two reactors.

The Gasifier's bed will consist of calcium oxide and calcium sulphate and will be fluidised with recycled gases. The wood feed, after drying, will be fed into the bed where oxygen from the calcium sulphate will oxidise or gasify the wood. The bulk of the recycled gas will act as carrier for the oxygen from the stone to the char.

In the Gasifier some heat will be liberated from the gasification or partial combustion of the wood, but the reduction of calcium sulphate is endothermic, so that the gasifier requires extra heat to maintain its temperature. The oxidation of calcium sulphide in the Oxidiser is exothermic, and this reaction will heat up the stone, enabling heat to be transferred with the stone back to the gasifier. When gasifying wood after allowing for heat losses and heat exchange between the incoming and outgoing gases there is a small deficiency of heat. This heat will be obtained by allowing a small portion of the char or carbon to leave the gasifier with the stone to be burnt in the oxidiser. The temperature differential required between the Gasifier and the Oxidiser will be determined by the quantity of stone being transferred between the beds. It should be noted that this system does not require extra excess steam to be passed through the Gasifier to keep it cool as is the case if pure oxygen or enriched air is used.

The wood feed will be cut and shredded and then partially dried using surplus heat from the reactors. The optimum dryness of the wood will be determined during the pilot plant trials. The wood feed will be fed and metered through a lock hopper into the Gasifier.

The Gasifier and Oxidiser will be of rectangular shape and integrated into a single unit to minimise heat losses, and to provide common straight walls to simplify the transfer of stone between the beds. This is shown in Figure II, and has a patent pending. (9) Slots at about 45° permit the stone to pass from one bed to the other. The fluidising gas will be reduced near the outlet slot, so that the bed will slump into the slot. Increased fluidising gas at the exit of the slot will help the material to flow through the slot. A more complicated system has been used successfully for 15 years on the pilot plant and 20 MW scale. This new system has been demonstrated on a small cold model, and it will soon be tested on a larger scale at Birmingham University as part of a study of the movement of fluidised beds.

The gases leaving the Gasifier will be passed through Primary and Secondary Cyclones to remove particulates from the gas. The particulate stream from the Primary Cyclone will be returned to the hottest part of the Gasifier, and the Secondary stream will probably be fed into the Oxidiser.

The gases leaving the cyclones will be first heat exchanged with the incoming recycle gas and then will be further cooled in a Waste Heat Boiler raising steam at a controlled pressure which could be up to 20 bars, before the gases are finally cooled in a direct water spray. This final cooler will remove the particulates too small to be caught in the cyclones. The water from the spray will be settled, cooled and filtered before being recycled to the water spray. It may be necessary to remove a small bleed of water from this system to prevent build-up of undesirable materials.

The cooled gas will be freed as far as possible from entrained water by a cyclone and mesh demister. Part of this gas will be recycled by a fan back to the Gasifier to provide the basic fluidising gas, the remainder will be the product synthesis gas, which may need further cooling to reduce the water content.

Air will be compressed, preheated and used to fluidise the 'stone' in the Oxidiser, and to oxidise the calcium sulphide back to sulphate. The flue gases leaving the Oxidiser will be passed through cyclones, before being used to preheat the inlet air, finally these flue gases will be used to dry the wood feed. The particulates recovered in the cyclones will either be returned to the Oxidiser or removed from the system.

The ash content of wood is low, usually between 0.2 and 3 percent, with a few woods up to 8 percent. (10) The bulk of the ash in wood is calcium oxide frequently 50% or more, and this obviously cannot react with the bed material, and so will not impose any temperature limitations on the process. It is expected that the major part of the ash will be recovered from the secondary cyclones; that from the Gasifier could be passed through the Oxidiser, and the Oxidisers' could be removed in all or part from the system. If any ash remains in the bed it will be allowed to build-up, and an occasional bleed-off of stone from the Oxidiser will be used to control its quantity.

There is a possibility that some of the calcium sulphate in the Oxidiser might decompose to contaminate the flue gases with sulphur dioxide. Excess air in the Oxidiser and lower temperatures would diminish this possibility, and experimental results show that the SO_2 content will be low, at most some hundred parts per million. Pilot plant operation will monitor this possible effluent, and the operating conditions required to minimise it will be determined. If it is at an unacceptable level then there are at least two ways of further reducing it. The flue gases leaving the Oxidiser could be passed through a bed of lime kept at 860°C which should remove about 90% of the SO_2.

Another possibility is to remove the secondary Cyclone from the flue gas, and permit some particulates to be caught in the wood drier. The moisture would slake the lime, and the calcium hydroxide formed is very reactive to sulphur dioxide and should again remove a substantial proportion of the sulphur in the flue gas, (11) returning it to the gasifier.

MATERIAL BALANCE

IN	Wood	900	Kg.
	Water	100	
	Air	2656.5	
	Fresh Stone	9.7	
	Process Water	112	
	B.F.W.	676	
		3778.2	Kg.

OUT	Synthesis Gas	1162.4	Kg.
	Flue Gases	2492.3	
	Ash	4.5	
	Reject Stone	10	
	Water Purge	109	
	Steam	676	
		3778.2	Kg.

MASS EFFICIENCY

0.019 Kmol of CO per Kg of dry wood from wood of 20% moisture

0.027 Kmol of H_2 per Kg of dry wood from wood of 20% moisture

The dry gas efficiency is over 75% as calculated as the Higher Calorific Value of the Synthesis Gas compared to that of the dry wood feed.

HEAT BALANCE

Basis - 1000 Kg of wood containing 100 Kg of moisture.

IN	Wood	50500 KCal.
	Wood	CV 3862300
	Air	22940
	B.F.W.	33800
	Fresh Stone	40
	Solution Heat NH_3	6
	Process Water	1680
	Cooling Water	129010
		4100276
OUT	Synthesis Gas	102300
	Synthesis Gas	CV 2898030
	Flue Gases	350360 *
	Steam	446800 *
	Ash	1370
	Reject Stone	2936
	CaO $\longrightarrow$ CaS	120
	Water Purge	4760
	Cooling Water	162000
	Heat Losses	131600
		4100276

* Heat available for drying wood.

NET INPUT

The input of oxygen to the process is nil as the necessary oxygen is obtained from air, by the application of the Oxygen Donor principle.

The process does not use any external steam. The process produces steam from the waste heat at the rate of 0.75 Kg of 7.9 bara steam if used for wood drying. This steam could be produced at pressures up to 20 bara if required for power production.

The process produces sufficient waste heat in the form of steam and flue gases to dry the wood feed from 40 to 45% moisture down to 10% which is considered to be a reasonable moisture for the wood feed to the plant.

Preliminary calculations indicate that the electric pumps, fans, conveyors, grinders, etc., necessary for this process will consume about 60 kWh per tonne of dry wood.

METHANOL PRODUCTION

The Synthesis Gas as shown on the material balance will have been scrubbed for removal of particulates, but will need further cooling to reduce the water content.

All the experimental evidence to date indicates that the methane content will be sufficiently low, not to warrant a demethanisation unit, and that it will only need a reasonably sized purge in the methanol synthesis unit.

Similarly, experimental evidence has shown the formation of tars to be low and unlikely to be produced in any quantity.

The synthesis gas will need compressing, the Shift Reaction to adjust the CO/H_2 ratio, followed by CO_2 removal and a guard to remove any final traces of H_2S before the gas can be fed to a methanol synthesis unit. It has been estimated that using the Oxygen Donor Process approximately two tons of dry wood will be required to produce one ton of methanol.

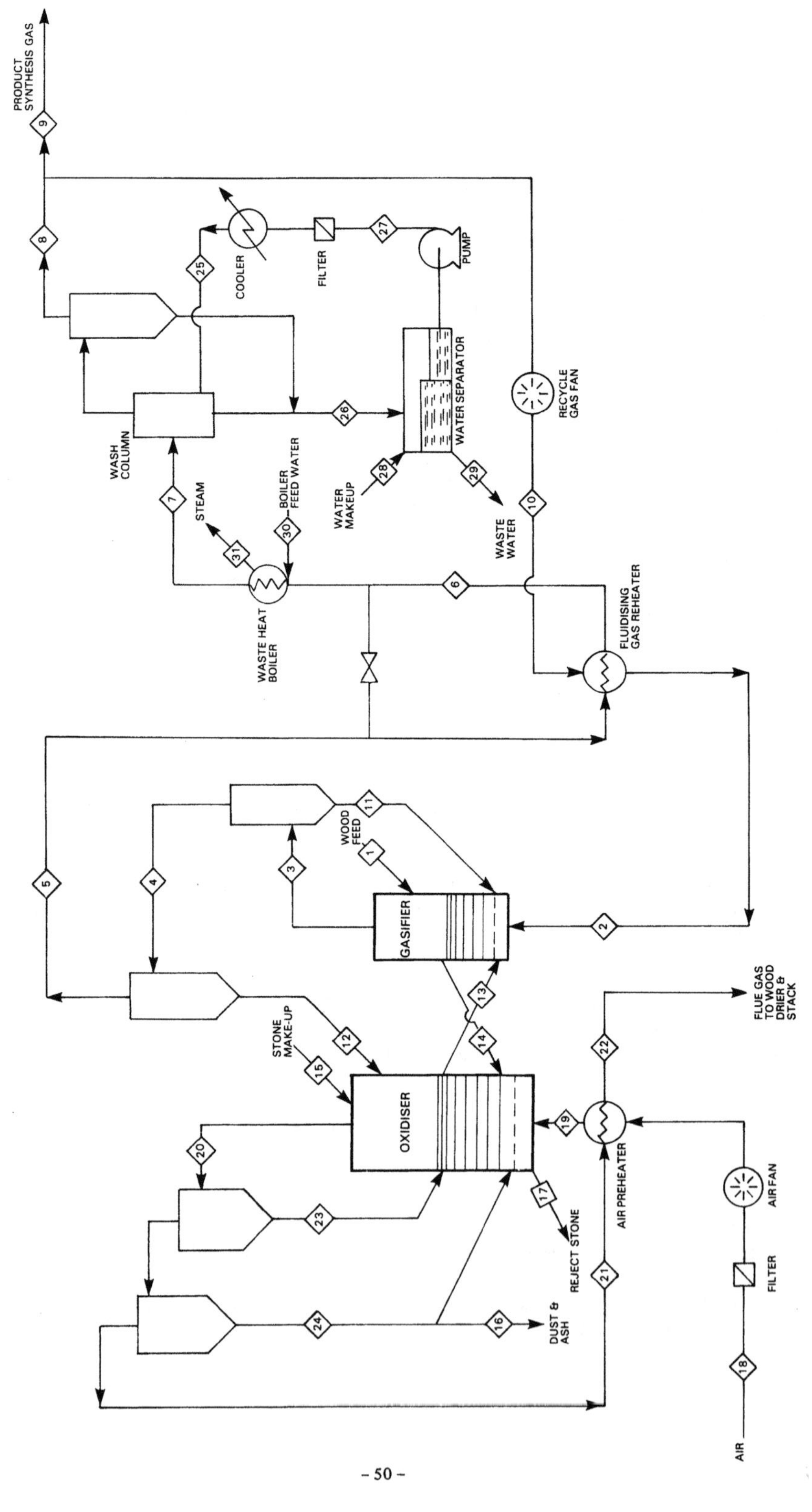

PRODUCT SYNTHESIS GAS
COOLER
FILTER
PUMP
WASH COLUMN
WATER SEPARATOR
RECYCLE GAS FAN
STEAM
BOILER FEED WATER
WATER MAKEUP
WASTE WATER
WASTE HEAT BOILER
FLUIDISING GAS REHEATER
WOOD FEED
GASIFIER
STONE MAKE-UP
OXIDISER
FLUE GAS TO WOOD DRIER & STACK
AIR PREHEATER
REJECT STONE
DUST & ASH
AIR FAN
FILTER
AIR

OXYGEN DONOR GASIFIER

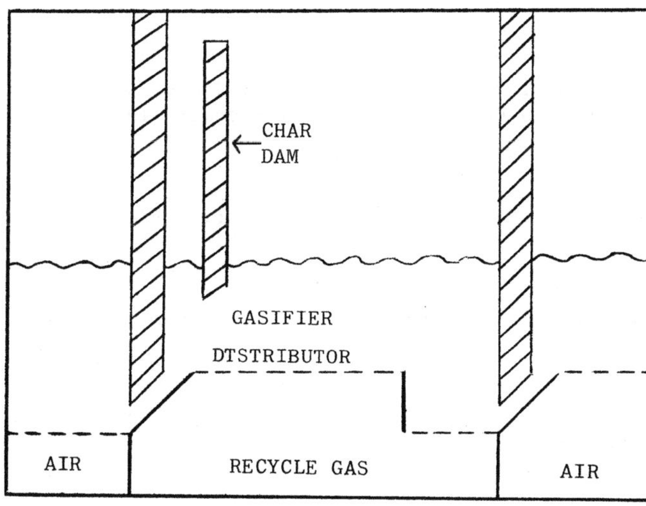

ELEVATION

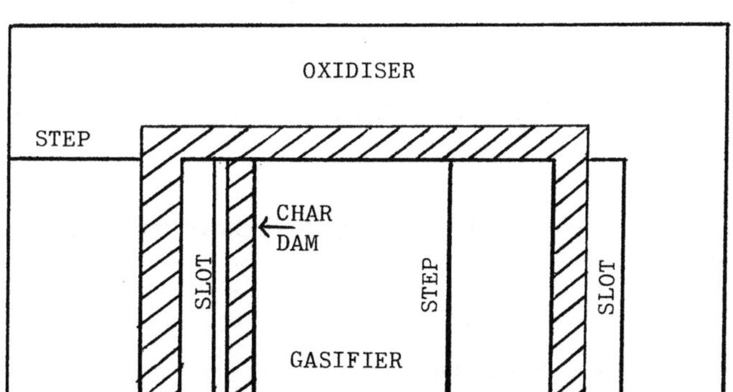

PLAN

REFERENCES

1. G.Moss "The Desulphurisation of Fuel Oil in Fluidised Beds of Lime Particles" United Nations Working Party on Air Pollution. Geneva, November 1970.

2. G.Moss, J.W.T.Graig & D.Tisdall. A. I. Ch. E Symposium Series, Vol. 68, No 126 pages 277 - 282.

3. A.W.Ramsden, Z.Kowszum. U.S. Dept Commerce. Natl. Tech. Inf. Serv. Rep. N PB - 29B 226/2GA, February 1979.

4. S.L.Rakes. Oil and Gas Journal, April 1980, 78 No. 16 page 31.

5. G.L.Johnes, S.L.Rakes, Pet. Inf. 22nd May 1980 48 pages 51 - 52.

6. First Trial of C.A.F.B. Pilot Plant on Coal. EPA -600/7-77-027 March 1977.

7. EPA 600/7-79-066 February 1977, Appendix C.

8. Patent U.S. 7900761 9th January 1979, European 13590.

9. Patent Application U.S. 7941220, 29th November 1979.

10. Louis E. Wise. Wood Chemistry Rheinhold Publishing Company, New York 1946.

11. V.F.Estcourt, R.O.M.Grutle, D.C.Gehri & H.J. Peters. Combustion. November 1978 pages 36 - 41.

SYNTHETIC FUEL FROM WOOD USING STEAM AND AIR

Authors	:	R.E. Holmes, D.F. Gibbs, R.S. Davis
Proposal number	:	ESE/9/009/UK
Duration	:	22 months 1 January 1982 - 30 November 1983
Total budget	:	Ƀ1,071,000 CEC contribution: Ƀ600,000 ECU
Head of project	:	Dr. R.E. Holmes Pollution Prevention (Consultants) Ltd.
Contractor	:	Pollution Prevention (Consultants) Ltd.
Address	:	Pollution Prevention (Consultants) Ltd. Crown House Copthorne Bank Crawley Sussex RH10 3JG England

Summary

Pollution Prevention Consultants, Inc. (PPC) is using existing fluidised-bed gasification technology to develope a synthetic fuel from wood process whereby the synfuel product is acceptable for use in a standard methanol-producing plant. The basis for this technology is the operation of a pilot plant and a demonstration plant which have provided a database in both the gasification and combustion modes. A brief description of the pilot plant units preceeds the discussion of the synthetic fuel from wood process.

FLUIDISED-BED GASIFICATION SYSTEM

The disposal of various agricultural, industrial, and municipal solid wastes in an environmentally acceptable manner is a serious problem. The fluidised-bed system provides a pollution-free process that converts these wastes into clean energy sources while significantly reducing the volume of waste material.

Solid wastes are converted into clean energy products consisting of low-heating-value gas, pyrolitic oil, and a char/ash mixture. Up to 80 per cent of the energy value of most materials is recovered in the form of these useful fuels. The fuels can be burned directly in conventional oil and gas burners, stored for later use, or sold as feedstocks, turning a waste disposal problem into an economic asset. A 20-ton-per-day test programme unit is in operation which has utilised feedstocks such as wood chips, sawdust, logging wastes, cotton gin trash, rice hulls, paper, peat, corncobs, sludges, polyethylene, waste oil, municipal solid waste, peanut shells, bagasse, coffee grounds, coal/waste mixtures, manure, and tyres. The relative quantities of gas, oil, and char produced are dependent on the temperature of the reaction. The quality of the gas produced is a function of the fluidising medium. If air is used as the oxidant, the product gas is low-heating value (620-2225 kcal/scm); if oxygen or steam is used, the product gas is medium-heating-value (2225-2560 kcal/scm).

Pyrolytic oil is a heavy black oil, similar to No.6 residual oil. The oil is oxygenated and has a heating value of 5300 to 7300 kcal/kg. High reaction temperatures result in further cracking of the heavy tars and a lighter fuel oil is produced. Its consistency varies from that of paint to that of a light asphalt at ambient temperatures. This product can be blended with residual oil, or it may be fired separately into either oil- or coal-burning facilities. Commonly, the pyro oil is not separated from the low-heating-value gas, but is fed directly into the boiler with the product gas.

The char is a fluffy, fine, powdered material which is low in volatiles and has an average heating value of 5800 kcal/kg. The entire ash content of the feedstock is contained in the char ash mixture which is separated from the hot gas stream by a set of high-efficiency cyclones. The volatile content of the char can be varied for a particular feedstock by changing the reaction temperature in the fluidised bed. The ash content of the char varies directly with the ash content of the feedstock. The char product can be blended with coal or oil and burned in conventional burners, or it can be used to produce a superior charcoal briquet.

The low-heating-value gas has a heating value between 620 and 2225 kcal/scm. Gas composition varies greatly with reaction temperature, fluidising medium, and the moisture content of the feedstock. The low-heating-value gas can be burned in a low-heating-value gas burner to produce hot air for drying or similar operations; it can be burned in a retrofitted boiler to be used for steam generation; or it can be used to generate electricity.

The pyrolysis/gasification system includes feed handling and preparation equipment, a dryer, reactor, and char collection equipment.

PILOT PLANT CAPABILITIES

PPC, through a licencing arrangement with ERCO, has been operating a 0.19 m^2 fluidised-bed pilot plant for 3 years. The present system is a combination pyrolysis/combustion fluidised bed, having removable cooling tubes in the fluidised bed.

The pyrolysis/gasification system includes a shredder, the reactor, start-up burner, mechanical particulate collection device, adjustable venturi oil removal system, and an afterburner. The system is rated as a nominal 22,000 kg/day of agricultural feedstock in the pyrolysis/gasification mode. The shredder is capable of sizing material to 0.6 cm. Several feeding arrangements are available to accommodate the wide variety of acceptable feedstocks. A 12.7 cm diameter screw feeder capable of delivering 6 m^3/hr is used to feed materials with a maximum moisture content of 30 percent. A rotary airlock above the screw seals the feed hopper against the slight positive pressure of the bed.

A second available feed system is the transition transport feeder, a pneumatic transport system that allows solids to be fed into the bottom of the bed. It is used primarily for feeding sorbents when the reactor is in combustion mode. Combinations of the aforementioned feed systems are also available.

The reactor itself has a 0.19 m^2 cross-sectional area, and is 5.2 m high. The bed consists of 0.11 m^3 of sand sized to 800 microns. The reactor is designed to operate between 400° C and 1100° C, with a maximum freeboard pressure of 0.15 atm. The fluidisation blower has a maximum capacity of 17 m^3/min at 0.5 atm. Fluidisation velocities range from 0.2 m/sec to 4 m/sec.

Particulate separation from the hot gas stream is accomplished by the use of one primary and three secondary high-efficiency cyclones. The fractional efficiency of the high-efficiency cyclones is 96 percent at 7 microns.

The oil scrubber is a venturi type scrubber. The maximum inlet temperature to the scrubber is 425° C with a gas exit temperature of 75° C. The oil scrubber is 99.7 percent efficient in particulate removal.

The hot gases are then combusted in an afterburner. The afterburner has its own combustion air blower capable of delivering 85 scmm. The afterburner is rated for 1.8×10^6 kcal/hr and is operated with a typical flame temperature of 1100° C. The outlet gases are cooled by direct water spray at the exit of the chamber.

FLUIDISED BED COMBUSTION

Pollution Prevention (Consultants) Ltd (PPC) has been engaged in fluidised-bed combustion work for the past 9 years. The 0.19 m^2 pilot plant fluidised bed has been used for combustion of coal, petroleum coke, and mixtures of refuse-derived fuel-coal and also for pyrolysis of various agricultural, municipal, and industrial wastes. We have recently started up our 3.4 m^2 (1.2 m x 2.7 m) fluidised-bed demonstration unit which is operated in the combustion mode and generates 9000 kg/hr of steam. The demonstration unit also includes an electrified bed which removes 99.5 percent of all gas particulates in the range of 40 microns down to 0.1 micron and a sand bed filter which removes 99.6 percent of gas particulates in a similar range. PPC's licensor, ERCO, has developed a proprietary system for feeding coal and limestone mixtures under the bed; the system has been successfully tested in a coal flow model with coal moisture contents as high as 5 percent.

PILOT PLANT (0.19 m^2 UNIT - COMBUSTION MODE)

Operation of the 0.19 m^2 fluidised-bed pilot plant (Figure 1) in the combustion mode requires the placement of heat transfer surfaces in the fluidised bed. A total of 40 transfer surfaces, consisting of inconel tubes (4.9 cm O.D.) can be placed in the fluidised bed. The tubes can be varied such that the distance between the distributor plate and the first tube bank can range from a minimum of 38 cm to a maximum of 132 cm. The horizontal placement can also be varied. The feed mechanism is a pneumatic underfeed system, in which the fuel and sorbent are fed together into the bottom of the bed. The pneumatic system gives better fuel distribution near the distributor plate. Additional versatility for combustion testing is provided by gas injection ports above the fluidised bed. These various ports are used for secondary air injection for two-stage combustion or injection for nitric oxide reducing agents such as ammonia.

Additional equipment particular to the combustion mode is an external shell and the tube heat exchanger which cools the combustion gas to protect the cyclone from the excessive temperatures of fluidised-bed combustion. In addition, a patented electrified bed (EFB) particulate control device removes the fine combustion fly-ash from the flue gas.

FBC DEMONSTRATION UNIT (3.4 m^2)

The 1.2 m x 2.7 m FBC Demonstration Unit, currently in operation, is designed for a maximum steam generation rate of 9000 kg/hr. It is of sufficient size to generate meaningful experimental data that are applicable to larger commercial units. A process flow diagram of the 1.2 m x 2.7 m is shown in Figure 2.

The boiler may be fired with a variety of coals containing as much as 4.5 percent sulphur and will meet the USA federal standards for SO_x, NO_x, and particulate emission.

The emphasis in the design has been to produce a unit which is not complex in mechanical equipment, yet sufficiently flexible and well-instrumented to generate meaningful experimental data. The unit will be used to test components and features that would be of critical importance in a commercial unit. Among the areas to be tested are the following: (1) tube materials, (2) solids feeding system, (3) effect of operating parameters on bed performance, (4) controllability and stability of fluidised bed, and (5) performance of a sand bed filter and an electrified bed for secondary particulate collection.

The FBC facility is a completely contained unit. It consists of subsystems for solids handling, the fluidised-bed boiler itself, air supply, waste solids, handling, and gas cleanup.

PILOT PLANT ACTIVITIES

The pilot plant facility is a highly flexible system. Not only is it able to accommodate a wide variety of feedstocks, but also it is able to operate in a variety of modes. In addition to combustion and gasification using air as the fluidising medium, the pilot plant is also equipped for true pyrolysis (where the heat necessary for pyrolysis comes from a source other than the feedstock) and steam gasification.

The present-day function of the pilot plant facility is to provide PPC with design information for commercial plants. In the past 18 months the design data for the gasification plants for the Kingsford Company and the Tenneco Company have been obtained. Adequate feed systems, solids separation, and maintenance of stable afterburner combustion have been the focus of these studies.

THE KINGSFORD SYSTEM

The Kingsford Company has licensed PPC technology to design and build a fluid-bed gasification system to produce charcoal quality char, and 8 x 10^6 kcal/hr of hot gases to be used for briquet drying.

There are three main characteristics that are required to produce acceptable char. The char should have a volatility on a dry basis of 19 $\pm$ 3 percent. The char must contain less than 20 percent ash and the particle size should be large enough to ensure the ability to make good briquets. The PPC fluid-bed process meets all three criteria for charcoal production.

The volatility of the char is easily controlled by varying the reaction temperature. The volatility can be kept constant by maintaining a steady bed temperature. The fluidised-bed reactor using an inert bed material can be easily kept within a temperature range of $\pm 5^{\circ}$ C. The particle size of the char can be varied by changing the fluidisation velocity. Higher fluidisation velocities will produce larger particle sizes. Acceptable particle sizes are easily obtained from the fluidised-bed system. The ash content of the char is probably the most difficult parameter to adjust; however, feedstock with less than 4 percent ash on a dry basis will produce an acceptable ash level.

The Kingsford system consists of 4 main subsystems: feeder and shredder system, dryer system, reactor and particulate removal system, and afterburner system. It is designed to handle 900 kg/hr of wet woodwaste, 50 percent moisture content. Woodwaste consisting of bark, sawdust, wood chips, and post peel are sized to 0.6 cm minus and then dried to 25 percent moisture content in a rotary drier. The 25 percent moisture content woodwaste is then fed by means of a screw conveyor into a fluid-bed gasifier. The vessel has an inner diameter of 1.8 m and is 6.1 m high. Air is used as the fluidising medium, and the bed temperature is 450° C. The reactor operates in pyrolysis mode using 20 percent of its feed to provide the heat for the pyrolysis reaction. Low-heating-value gas, oil, and char are produced. The char is separated from the warm gas stream by a system of high-efficiency cyclones.

The char is cooled by direct contact water spraying. The low-heating-value gas and pyrolytic oil are combined in the afterburner along with any entrained particulate. The hot gases exciting the afterburner are then cooled to 315° C with diluent air. Twenty-five percent of the 315° C gas is used to dry the feedstock to 25 percent moisture content. The remaining 75 percent of the hot gas produced is used for briquet drying. A process flow diagram for the Kingsford system is shown in Figure 3. The two most important features of the Kingsford system are that there are no wastes generated and that the system is 93 percent energy efficient.

The previous applications of fluidised-bed pyrolysis to a variety of feedstocks shows the wide range of use of the process. PPC is using its knowledge in the area of fluidised-bed technology to expand the European energy supply via the conversion of waste products to usable forms of energy.

PROCESS DESCRIPTION FOR PPC SYNTHETIC FUEL FROM WOOD USING STEAM AND AIR

The process of converting wood material to methanol is a problem suitably handled by fluidised-bed gasification technology. PPC has made an arrangement to licence the existing ERCO technology. PPC will use this technology as a basis for developing the syngas from wood technology required by the EEC. PPC will be the prime source of the technology and will be assisted by International Research and Development Co Ltd. The proposed plant will be fabricated in the U.K. by Northern Engineering Ltd. and installed by Hildebrand Co in Germany for our client Bayrl.

In this description the syngas to be utilised in the methanol plant will be burned in an internal-combustion engine to yield a net value of 800 Kw(e).

SYNGAS PRODUCTION

Approximately 750 kg/hr of raw wood, 50 percent moisture (Figure 4) is fed to a rotary dryer which reduces the moisture content to 20 percent. A major portion of this dry wood (stream 9) is fed to a fluidised-bed gasifier operating at 7.8 atm and 727° C. The fluidised-bed medium is high-temperature refractories, and the fluidising gas is high-temperature steam (stream 14). The gasifier generates 1446 scm/min of synthesis gas (stream 19) which has a higher heating value of 3030 kcal/scm (dry) or 2400 kcal/scm (wet); the synthesis gas composition (volume percent) is $CO/CO_2/H_2/CH_4/H_2O$: 18.8/15.9/35.2/9.4/20.7. Entrained particulate from the gasifier is removed and recycled to the gasifier via a cyclone. The heat required for the gasification reactions is supplied by recycling hot sand (stream 13) from a fluidised-bed combustor. The sand recycle ratio is nominally 12 to 1 (stream 13 divided by stream 9). Spent char (stream 12) from the gasifier is burned in the combustor to maintain the temperature at 982° C; the char yield from the wood gasification is nominally 10 percent.

Since the char rate is not sufficient to maintain the combustor at 982° C, additional wood (stream 11) is added to the combustor to provide the necessary heat. The hot combustion gases (stream 18) are passed through a gas expander to run the combustor fluidising air compressor. The exhaust gases (stream 2) provides the heat of vaporisation to the dryer. Particle-laden off gases are sent to a scrubber to remove environmental pollutants.

The circulation of solids between the two fluidised beds will be accomplished by utilising conventional solids circulation techniques which are presently demonstrated commercially in catalytic cracking processes. The flow of hot solids from the combustor to the gasifier will be maintained by gravity feed of fluidised solids from the combustor which is elevated relative to the gasifier. The flow of solids from the gasifier vessel to the combustor will be maintained with a less dense lift leg which utilises a portion of the combustion air to establish the density difference required for flow. A high-temperature slide valve will be provided in this circulation line to control the solids circulation rate which sets the gasifier temperature.

SYNGAS UPGRADING

The syngas produced in the gasifier reactor is sent to the syngas upgrading section (Figure 5). Here the syngas is processed so that the impurities are removed and the CO/H_2 ratio is at the level required for methanol production. A majority of the syngas produced (stream 20) is fed to an H_2O scrubber where the gas is cooled to 66°C with a water spray (stream 22) to condense out a large portion of the water vapour and tars present. The cooled gases (stream 26) are then heated to 815°C in heat exchanger HX-1 by indirect contact with hot syngas from the reformer.

The reformer is a nickel catalyst reactor in which the heat of reaction is provided by combustion of a portion of the syngas product, (stream 21 (21.3 per cent of stream 19)). Superheated steam (stream 28), heated in heat exchanger HX-3 by the reformer combustion gases, is added to the syngas (stream 27) in the reformer. At conditions of T=927°C, P=7.8 atm, the syngas reforms resulting in the percentage volume ratio of $CO/CO_2/H_2O/CH_4$:20.17/9.65/41.54/28.51/0.13. The H_2/CO ratio of stream 34 is 2.06, and the mass efficiency = 0.054 (kmol H_2+CO produced)/kg dry wood). The reformer product gas is cooled to 471°C by the syngas in HX-1 and to 234°C in HX-2 by providing heat of vaporisation for the gasifier fluidising steam (stream 14). The cooled gas (stream 37) enters the CO_2 remover where the CO_2 content is diminished.

The product gases (stream 39) could be further cooled to condense out the water present and thereby utilised in a methanol production plant. In this PPC design, stream 39 is fed to an internal combustion engine where the product gas is burned to provide:

1. 1012 KWe
2. 1.05×10^6 kcal/hr of 91°C hot water (stream 46)
3. 0.059 kcal/kg exhaust gas (stream 42).

The electricity provides power to operate the air compressors, blowers, and pumps with a resulting net electricity of 800 KWe. The exhaust gas is expanded through a turbine to run the internal combustion engine air compressor. The resultant exhaust gas (stream 43) can heat the engine cooling water (stream 46) to provide 2×10^6 kcal/hr of hot water at 100°C.

The requirements of the plant are 4.8 kg input of air/kg dry wood, 19.3 kg input of 21°C 1 atm water/kg dry wood, and 0.67 kw electricity input/kg dry wood to produce syngas only.

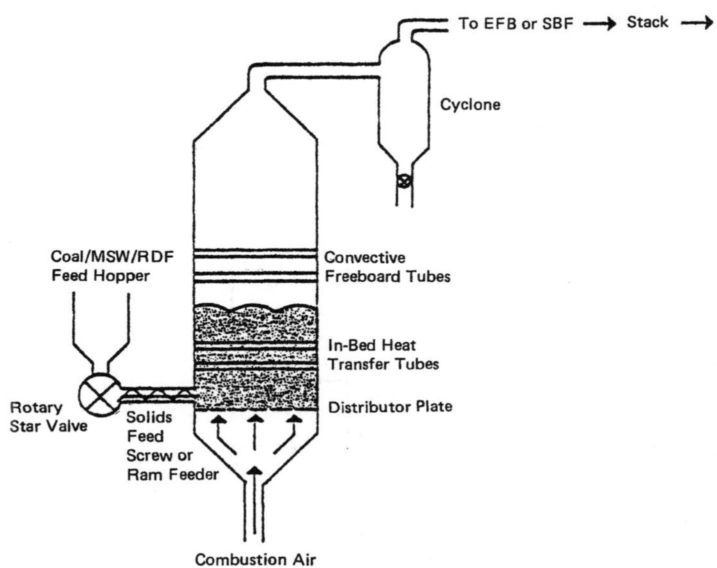

Figure 1. Schematic of 0.19 m^2 Fluidized Bed Pilot Plant.

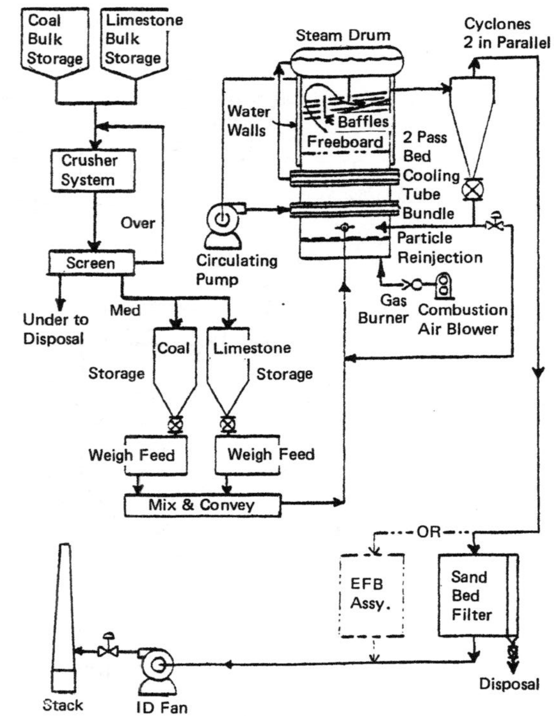

Figure 2. Process Flow for the 9,000 kg/hr Fluidized Bed Coal Combustion Plant.

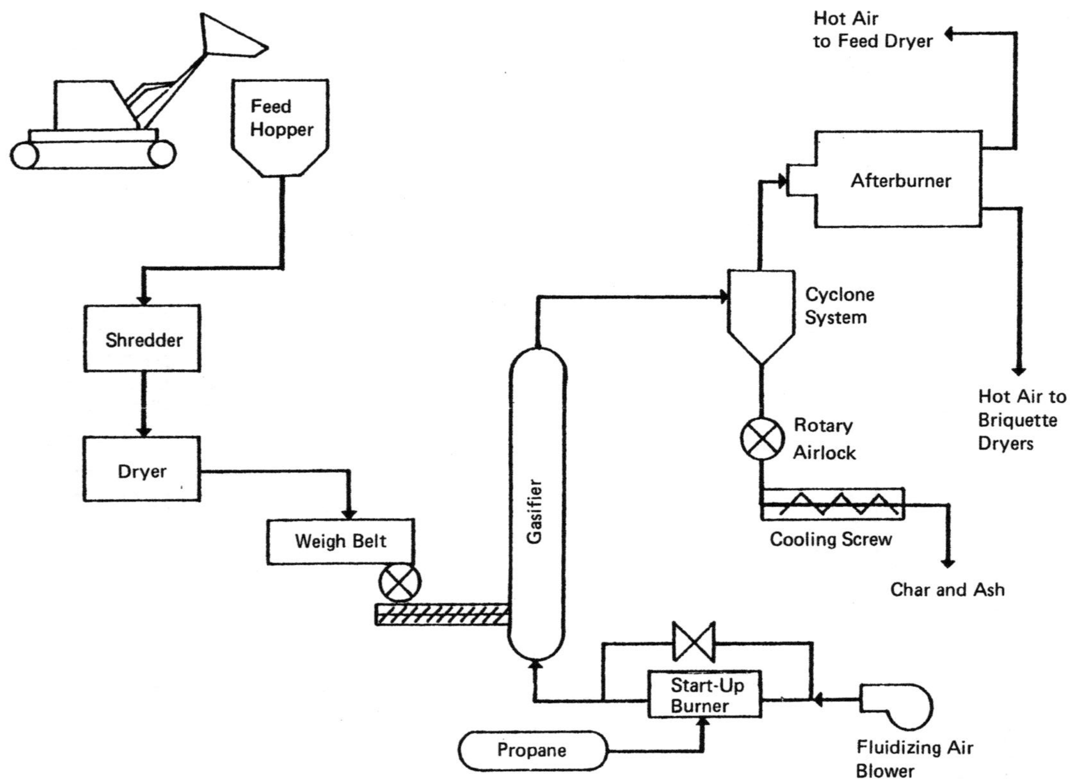

Figure 3. Process Flow for Fluidized Bed System for Wood-to-Char Gasification.

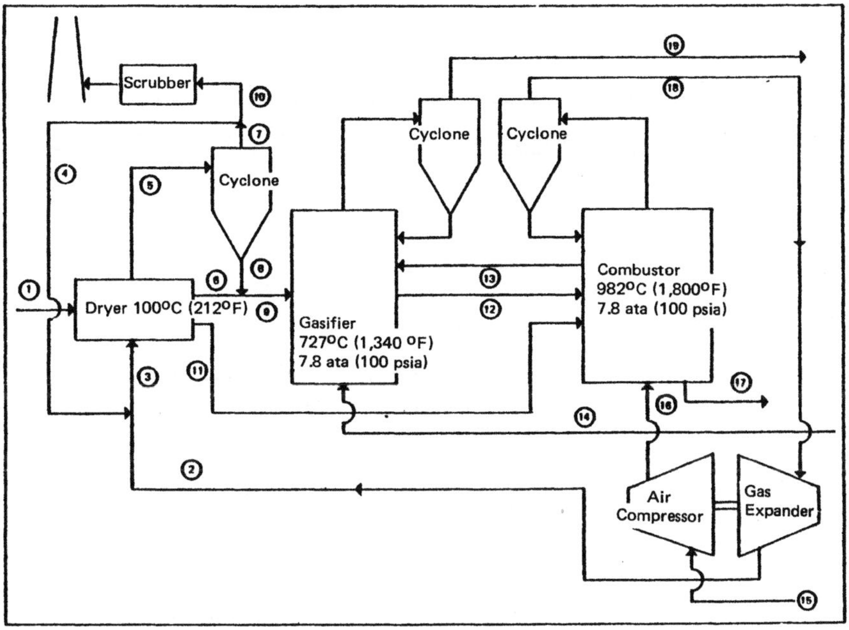

Figure 4. PPC Wood-Syngas Process (Syngas Production).

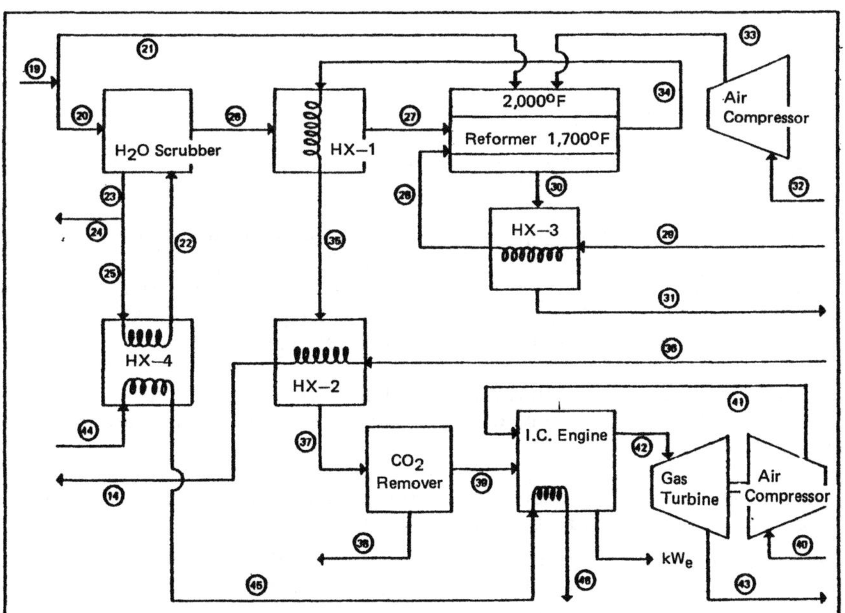

Figure 5. PPC Wood-Syngas Process (Syngas Upgrading).

EXPERIMENTAL WORK ON A FIXED-BED OXYGEN GASIFIER

IN THE VIEW OF METHANOL SYNTHESIS USING BIOMASS AS A FEEDSTOCK

Proposal number :	ESE - P - 010 - F
Expected duration :	2 years
Main contractor :	CREUSOT LOIRE ENTREPRISES (CLE) 33, Quai Galliéni 92150 SURESNES
Other contractors :	- Centre National du Machinisme Agricole, du Génie Rural, des Eaux et des Forêts (CEMAGREF) - Centre Technique Forestier Tropical (CTFT) - Société Nationale Elf-Aquitaine (SNEA) - Total Energie Développement (TED)
Head of Project :	Mr Michel COURTINAT, Directeur Energies et Techniques Nouvelles, Creusot-Loire Entreprises

Summary :

Methanol production requires a syngas with low content of methane and hydrocarbons, and no nitrogen.

As far as biomass gasification is concerned, methane and hydrocarbons come from pyrolysis products (50 % in energy and 80 % in mass). By mean of a gas recycling system, such pyrolysis products are fed into a combustion chamber together with oxygen, and eliminated at high temperature (1000-1100°C). Heat developed in the chamber is used for char gasification.

Temperature level in the combustion chamber and in the char bed is controlled by mean of the recycling flow so as to avoid ashes melting.

This recycling system has been experimented with oxygen for long periods on an horizontal fixed bed gasifier, at the rate of 1 t/h of dry wood.

Mass efficiency of this system is 38,7 kmole/kg of dry wood, and methane content is found below 2 % with wood at 30 % moisture.

The project is to raise gasifier pilot to an imput of 2,5 t/h at 10 bars operating pressure.

FRENCH JOINT-VENTURE " BIOMASS - METHANOL " :

CREUSOT - LOIRE ENTREPRISES ENGINEERING COMPANY

ELF AQUITAINE } OIL COMPANIES

TOTAL ENERGIE DEVELOPPEMENT }

C T F T CENTRE TECHNIQUE FORESTIER TROPICAL

C E M A G R E F

SET UP TWO YEARS AGO.

. TECHNO ECONOMICAL STUDIES ON THE PROCESS " WOOD TO METHANOL "

. TECHNOLOGICAL WORKS ON THE C E M A G R E F GASIFICATION TECHNOLOGY.

BIOMASS AS A FUEL

1 - LOW MELTING POINT OF ASHES : 900° C

2 -

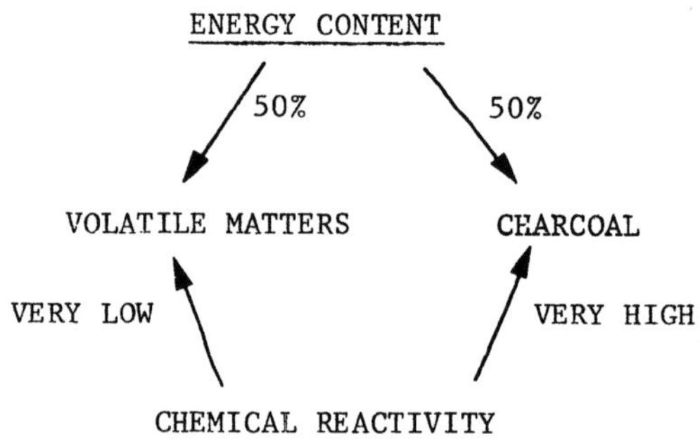

3 -

VALIDITY OF THEORETICAL CALCULATIONS BASED ON THERMODYNAMICAL EQUILIBRIUM.

GLOBAL ANALYSIS

METHANOL PRODUCTION REQUIREMENTS	GASIFIER CHARACTERISTICS	
	GENERAL	PARTICULAR
. SYNGAS (WITHOUT N_2)	OXYGEN BLOWN	COMBUSTION CHAMBER RECYCLING PROCESS
. LOW METHANE AND TARS CONTENT	HIGH TEMPERATURE TREATMENT OF THE PYROLYSIS PRODUCTS	" "
. EFFICIENCY	PRESSURE	10 BARS (cf model)
PREPARATION COST OF WOOD	MAXICHIPS	PACKED BED

VOLATILE MATTERS

(METHANE, TARS)

. 50% OF THE WOOD ENERGY CONTENT

. 80% ON A WEIGHT BASIS

. METHANE FROM PYROLYSIS : 10% IN THE PRODUCTED GAS

. PRESSURE ↗ ⟶ METHANE ↑

ELIMINATION OF METHANE AND TARS.

A. LOW TEMPERATURE GASIFICATION (800° C)

typical : fluid bed gasifier

methane content of the gas : relatively high

NEED FOR TREATMENT AFTER THE GASIFIER

" Post-combustion "

θ ↗ 1000 - 1100° C

ex Winkler process

pb : ash fusion

Reforming of the hot gas on catalyst

ex Eco process
Studvisk process

pb poisoning of catalyst

reforming after purification step

ex classical process.

pb energetic efficiency.

B. TREATMENT OF PYROLYSIS PRODUCTS AT A HIGH TEMPERATURE (1000 - 1100° C) INSIDE THE GASIFIER

. Despite the very low chemical reactivity, methane and tars desappear very quickly and completely at these temperatures.

. Very small reconstitution of methane

. producted gas close to thermodynamical equilibrium.
Model validity.

DELACOTTE AND C E M A G R E F PROCESS.

RECYCLING PROCESS

. "WASHING" OF PYROLYSIS PRODUCTS

. HANDLE THEM TO COMBUSTION CHAMBER (ELIMINATION OF METHANE AND TARS)

. LIMITATION OF TEMPERATURE EVEN WHEN USING O_2 (ASHES FUSION)

. IMPUT OF THE GASIFIER INCREASED (FAST PYROLYSIS AND DRYING)

RESULTS

	EXPERIMENTALLY ACHIEVED	EXPECTED FROM THE PILOTE PLANT
Gazification agent	pure oxygene	pure oxygene
Input (t/h dry basis)	1	2,5
Wood moisture (%)	15 - 45	10 - 15
Steam injection	0	0
Operating pressure (bars)	1	10
Feedstock size	maxichips small logs	maxichips
Oxygen input (kg/dry t)	350 - 400	380
Power input (kWh/t)		
Gas analysis (%)		
CO	31	40
H2	29,4	34
CO2	31,4	22
CH4	2,2	4
N2	6	0
Mass efficiency (Moles H2 + CO per kg of dry wood 20 % moisture)	38,7	46,4

CREUSOT-LOIRE ENTREPRISES	GAZOGENE EN LIT FIXE DU CNEEMA	Page 1

TRANSPOSITION POUR FONCTIONNEMENT A 10 BAR

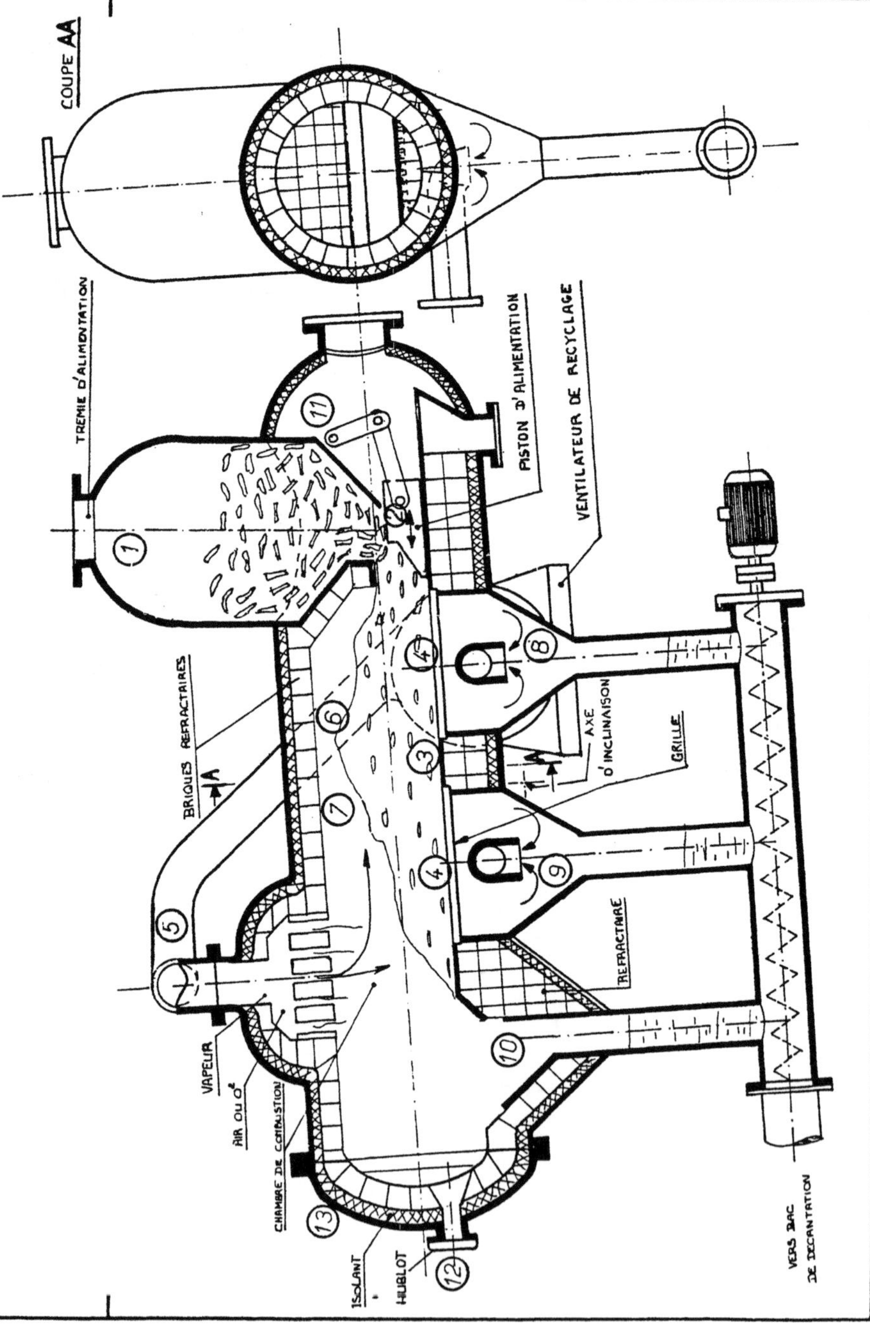

GAZOGENE EN LIT FIXE DU CNEEMA _(COUPE SCHEMATIQUE)

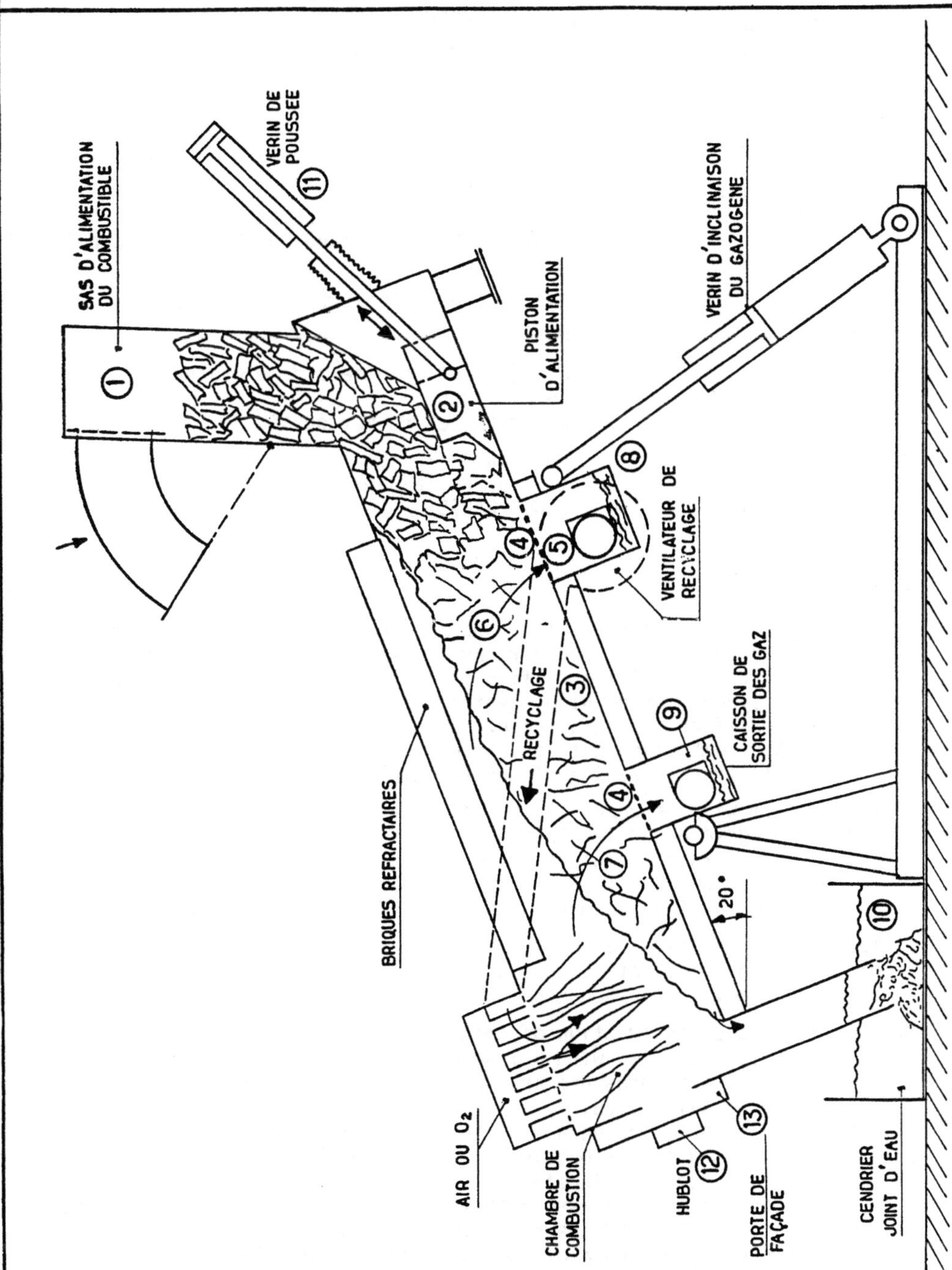

DESIGN AND CONSTRUCTION OF A PRESSURIZED WOOD GASIFIER WITH HEAT IN PUT BY OXYGEN COMBUSTION OR BY ELECTRICAL HEATING

Contracts Number	ESE-P-015/F
Duration	21 month January 1, 1982 - October 1, 1983
Head of scientifics studies :	Pierre DUBOIS Laboratoires de Marcoussis Route de Nozay - 91460 MARCOUSSIS (France)
Head of project :	M. HELARY Ateliers et Chantiers de Bretagne
Contractor :	NOVELERG Ateliers et Chantiers de Bretagne CGEE-Alsthom Laboratoires de Marcoussis
Adress :	NOVELERG 12, rue de la Baume 75008 PARIS

Summary

1. The purpose of the program is to design and build a gasifier with a capacity of 50 tons of wood per day. It will be a fixed bed down draft vertical gasifier operating at a pressure of 10 to 15 bars and producing syngas for methanol synthesis. The heat in put needed for charcoal gasification is obtained :
   - or by the combustion with oxygen of the recycled pyrolineous gases,
   - or by electrical heating of the recycled gases.

2. The program will be conducted by Novelerg and carried out by three other Compagnie Generale d'Electricite (CGE) subsidiaries :
   . Les Ateliers et Chantiers de Bretagne,
   . CGEE-Alsthom,
   . Les Laboratoires de Marcoussis, the CGE research center.

   Technip, which has a large experience in the design of plants producing methanol from natural gas will bring its support for the problems related to methanol production from syngas.

3. The design and construction of the gasifier will benefit from :
   - the results obtained with a laboratory apparatus : gas composition and influence of pressure and temperature on the kinetics of pyrolisis and gasification,
   - the results obtained on a gasifier of identical layout which will soon be tested at Marcoussis. This gasifier has a height of 13,5 meters, an internal diameter of .6 meter and an outside diameter of 1,3 meter. It will be used both for extensive test of the key components and for test of the complete system : influence of temperature and pressure, influence of the mass flow of recycled gases... on the quantity and composition of gas produced.

1. Companies involved

The companies involved in the program are 4 Compagnie Generale d'Electricite (CGE) subsidiaries. The program, conducted by Novelerg, will be carried out by the three following companies :
- Les Ateliers et Chantiers de Bretagne,
- CGEE-Alsthom,
- Les Laboratoires de Marcoussis, the CGE research center.

2. General layout of the gasifier

The gasifier which will be tested and built will be a fixed bed downdraft gasifier (cf. figure 1) working at a pressure of 10 to 15 bars and producing gases for methanol synthesis with a wood capacity of at least 50 tons per day. The pyrolisis and gasification zone will have a height of 11 meters and an internal diameter of 1,3. The heat input needed for charcoal gasification will be obtained :
- or by the combustion with oxygen of the recycled pyrolineous gases,
- or by electrical heating of the recycled gases.

The mass flow of recycled gases will be chosen to obtain the proper temperature in the gasification zone and avoid ashes melting. The fact that the recycled gases are brought to a high temperature by combustion or by electrical heating limits the methane content of the syngas produced and gets rid of the pyrolineous heavy products.

3. Time schedule

The program is a 21 month program, including three month for the gasifier installation on site and for tests leading to operation of the gasifier at full capacity.

4. Cost of electrical heating versus oxygen combustion

We have computed that the production of 1 kilo of syngas from dry wood with 20 % moisture requires :

- with oxygen combustion	2,5 kilo of wood 0,6 Nm^3 of oxygen
- with electrical heating	1,6 kilo of wood 2,5 kWh

It shows that oxygen combustion requires 50 % more wood than electrical heating and calls for a larger gasifier.

The comparison of raw materials costs shows that, with a price of 5 US cents per Nm^3 of oxygen :

. for a price of wood of 2 US cents per kilo which can be expected in countries with very large forests like Brazil, it is cheaper to use electricity if the price of a kWh is smaller than 2 US cents, which is the case in Brazil or in Canada where cheap hydroelectrical power is available,

. for a price of wood of 5 cents per kilo which can be expected in Europe, it is cheaper to use electricity if the price of a kWh is smaller than 3 cents, and that is the case in France for about 6000 hours per year.

5. Results backing the program

The program proposed to the Commission of the European Communities will be backed by results obtained on a laboratory apparatus and by results obtained with a test gasifier of significant size which is going to be tested at Marcoussis.

51. Test on a laboratory apparatus

We have designed and built a laboratory apparatus for pyrolisis and gasification studies. A pressurized vessel (cf. figure 2) with oxygen and steam inlets, with internal electrical heating and with good thermal insulation can contain one liter of wood or charcoal. This vessel can be operated up to 20 bars and 900°C. The amount of H, CO, CO_2 and CH_4 contained in the gas produced is measured by a chromatograph. Interesting results have been obtained showing :

- a strong influence of pressure on the pyrolisis kinetics. For the pyrolisis at 300°C of 8 cm^3 oak cubes, one can see on figure 3 that when pressure goes up from one to 6 bars, the time needed to produce 90 % of the gas, goes down from 38 minutes to 16 minutes. And, in this experiment, starting from room temperature, the time before any significant pyrolisis happens, goes down from about 45 minutes to about 12 minutes,
- a strong influence of temperature and a more limited influence of pressure on charcoal gasification kinetics (cf. figure 4),
- the composition of the gas obtained from wood pyrolisis or from charcoal gasification (cf. figure 5).

52. Results obtained with a test gasifier of significant size

We have designed a test gasifier of significant size with the general layout shown on figure 1. The main pressure vessel has a 1,3 meter external diameter and a 13,5 meter height. It is a fixed bed downdraft gasifier which can be operated up to 20 bars.

On the recycled gases loop, an electrical heater or a combustion chamber with steam and oxygen inlets can be used to bring in the gasifier the high temperature heat needed for wood gasification. The blower on the recycled gases loop will be driven by a variable speed electrical motor so that we can study the influence of recycled gases mass flow on the gasifier gas production, quantity and composition.

Special care must be taken of safety problems both at the design stage and during the test of gasifier, due to the combination of high pressure and high temperature operation and toxic and flammable gases. Special attention has been payed to wood feeding and ashes removal and to the combustion chamber.

Wood will be fed on a batch basis through a pressurized lock hopper (cf. figures 6 and 7). The lock interspaces will be alternatly pressurized with gas produced by the gasifier or filled with air after the gas has been sent to the flare. A vacuum pump will prevent any mixture of air and gas during the lock hopper operation.

The gasifier is now undergoing factory tests at Ateliers et Chantiers de Bretagne in Nantes and it will be soon be sent to Marcoussis for extensive tests of both the complete system and the key components.

It must be pointed out that this gasifier having an inside diameter .6 meter and an outside diameter of 1,3 meter versus an inside diameter of 1,3 meter for the larger gasifier, key components tested on this gasifier will be extrapolated with no significant problem for the larger gasifier.

6. Mass balance and wood burning capacity

Using the results obtained on the laboratory apparatus, we have computed wood burning capacity and mass balances :

- figures 8 and 9, show the mass balances computed for different values of the H_2/CO ratio and for different values of the ratio of water to minimum steam needed. It must be pointed out that for cost reasons, it should be better not to try to obtain in the gasifier the value of 2 for the H_2/CO ratio but to have shift reactions outside the gasifier,
- on figure 10, we compare the wood burning capacity computed of the Marcoussis gasifier and for the project. One can see that the computed value for the project is nearly 100 tons a day at 9 bars.

Figure 1

GASIFIER - GENERAL LAYOUT

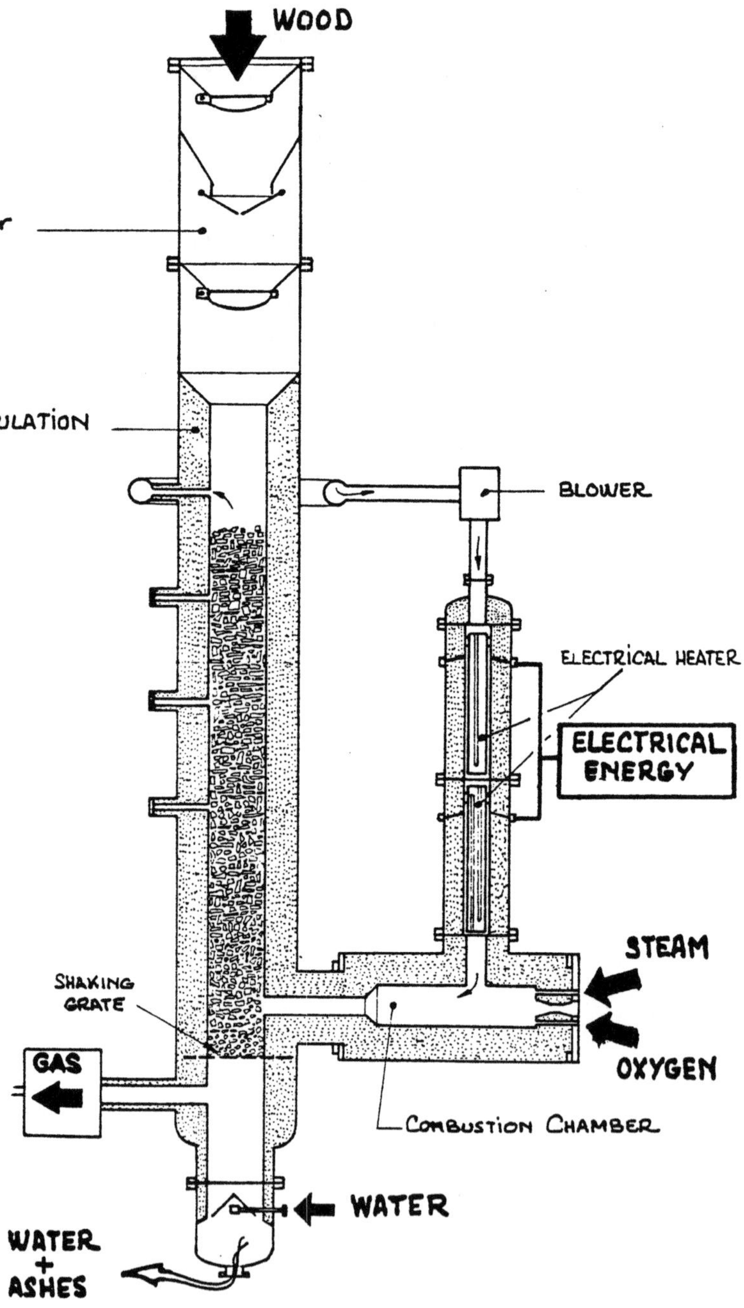

Figure 2

VESSEL FOR PYROLYSIS AND GASIFICATION STUDIES

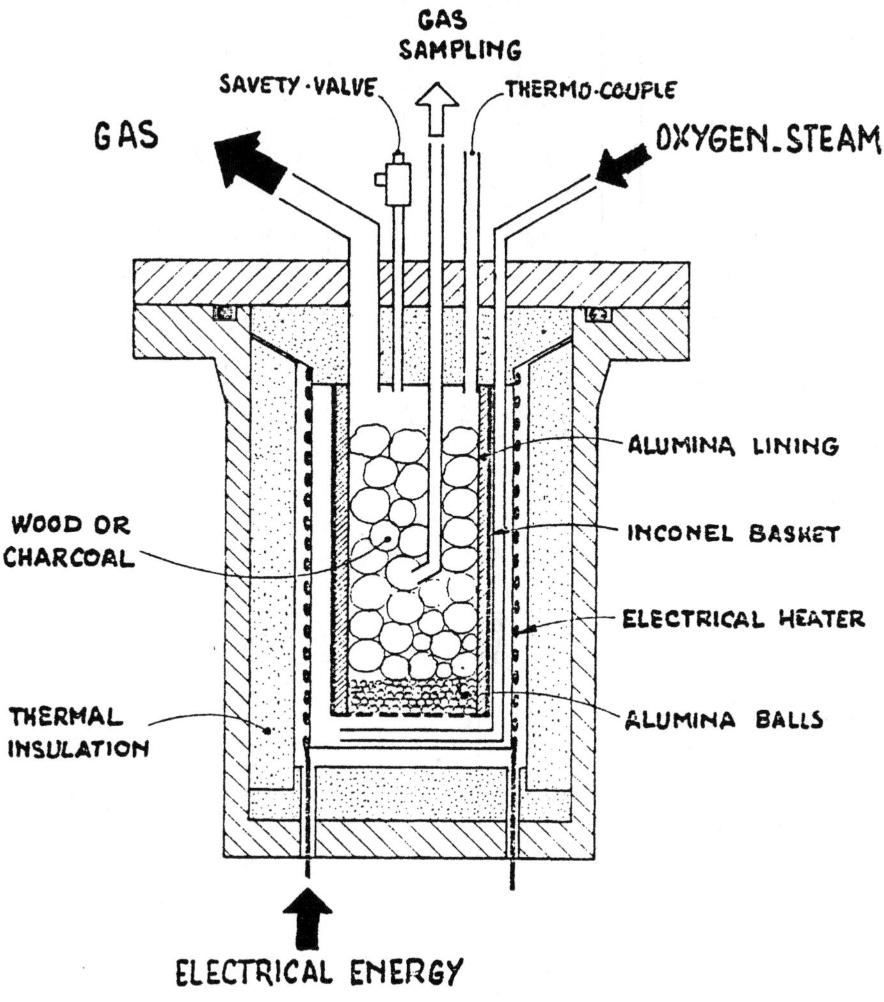

Figure 3

INFLUENCE OF PRESSURE ON THE PYROLYSIS AT 300°C OF $8 cm^3$ OAK CUBES

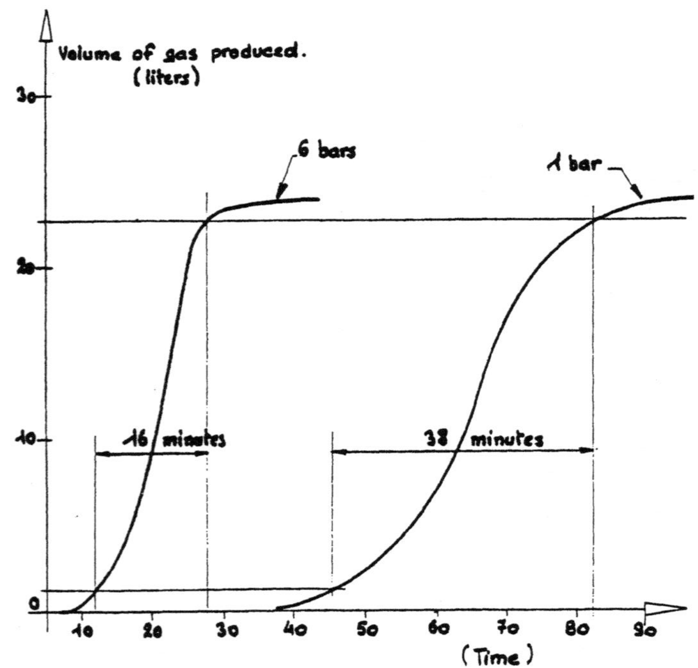

Figure 4

CHARCOAL STEAM GASIFICATION

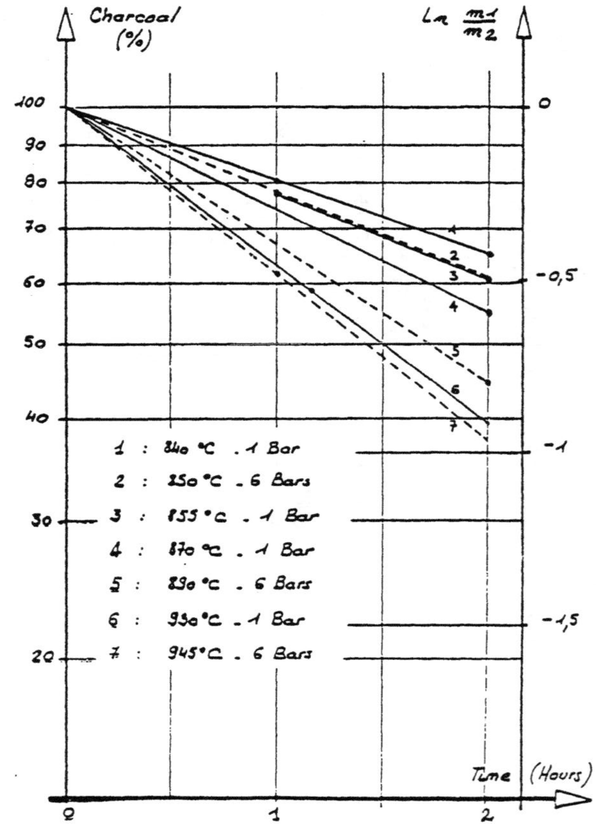

Figure 5

LABORATORY PYROLYSIS AND GASIFICATION STUDIES

I. WOOD PYROLYSIS (ELECTRICAL HEATING - 9 BARS)

GAS COMPOSITION (VOLUME %)

TEMPERATURE (°C)	H_2	CO	CH_4	CO_2
300	0	22 - 29	7 - 12	61 - 71
500	0	50 - 52	22 - 23	25 - 27
700	0,5 - 1,5	22 - 24	31 - 32	43 - 46
TOTAL	0,3 - 0,4	30 - 35	19 - 20	46 - 50

II. CHARCOAL STEAM GASIFICATION (930° - 1 BAR)

GAS COMPOSITION (VOLUME %)

H_2	CO	CH_4	CO_2
52	30	1,1	16

Figure 6

TEST GASIFIER - WOOD LOADING

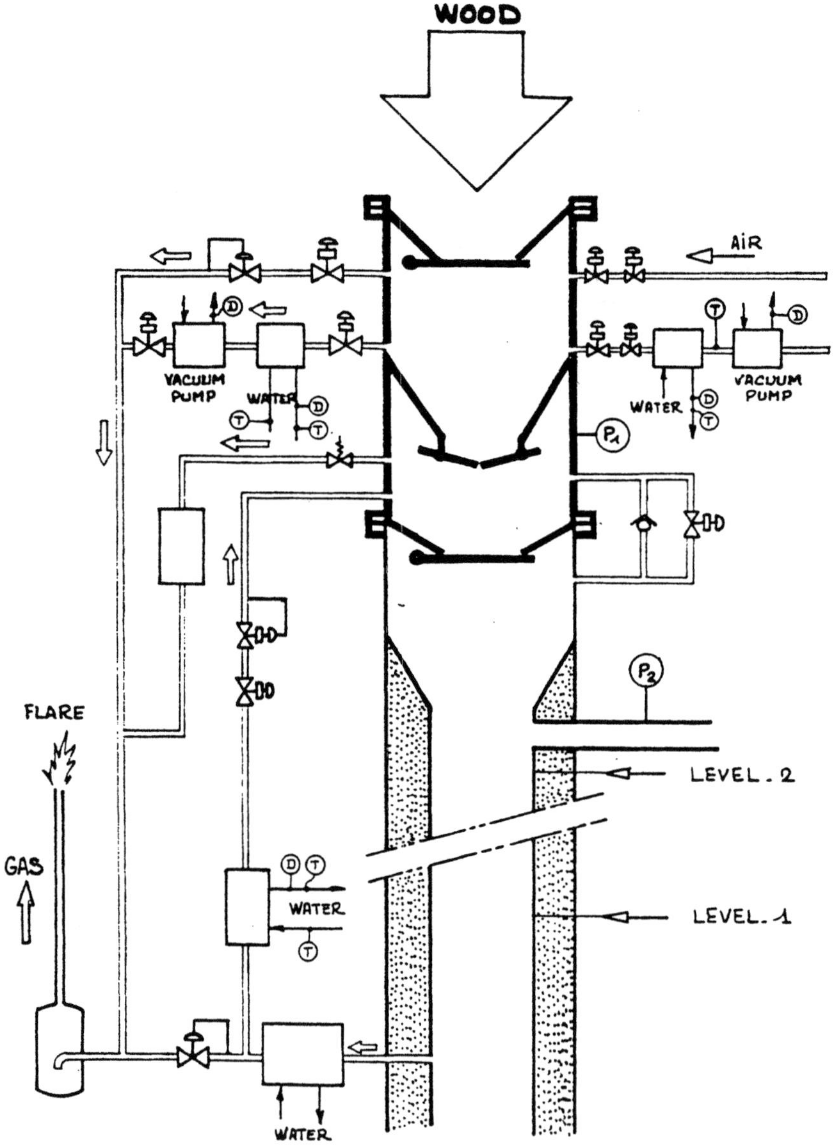

Figure 7

TEST GASIFIER

WOOD LOADING

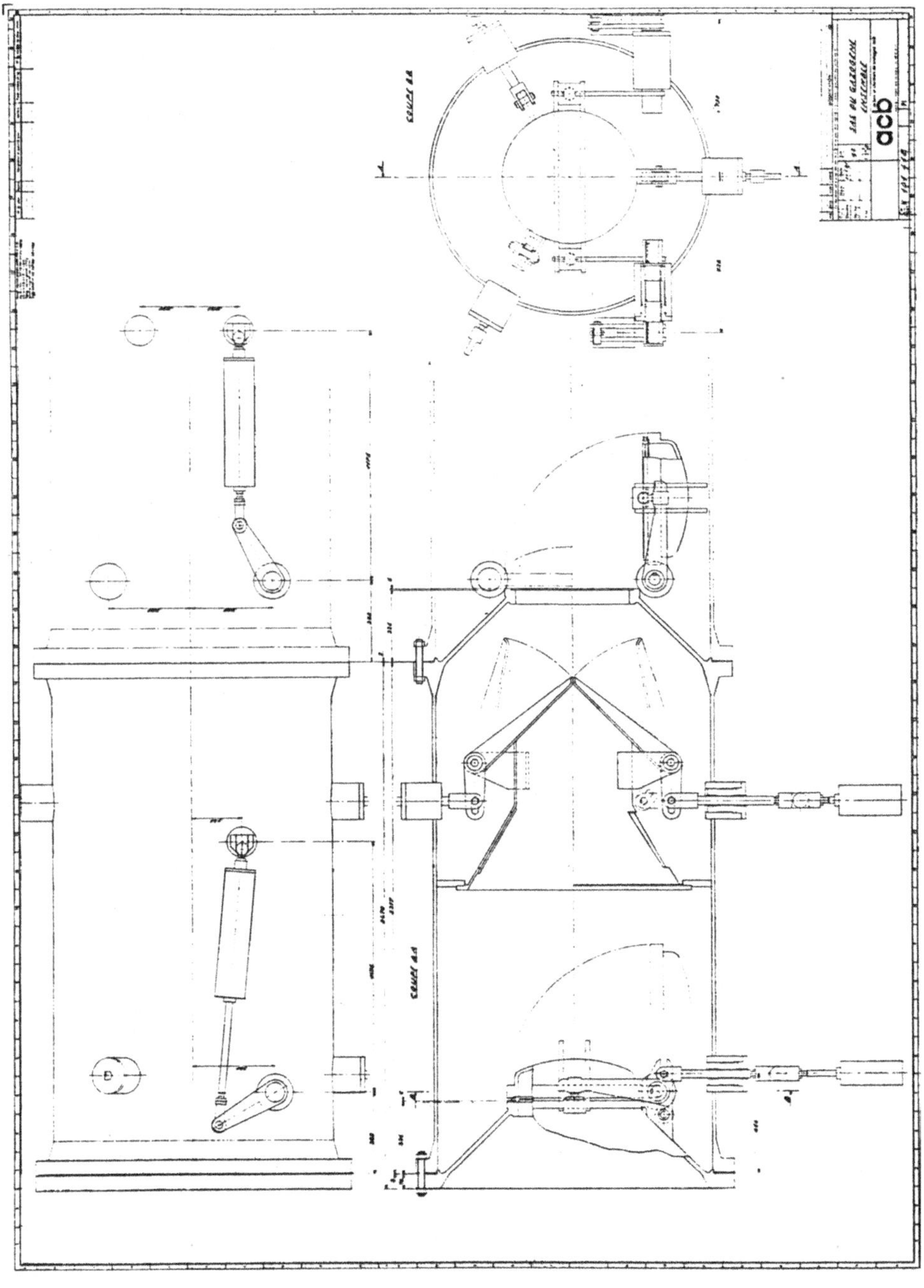

Figure 8

WOOD STEAM GASIFIER WITH ELECTRICAL HEATING

MASS BALANCE PER KG OF DRYWOOD
OF 20 % MOISTURE

	H_2/CO = 1,5		H_2/CO = 2	
INPUT				
- DRYWOOD OF 20 % MOISTURE				
. WOOD	0,8	0,8	0,8	0,8
. WATER	0,2	0,2	0,2	0,2
. STEAM	0,10	0,70	0,18	0,95
WATER + STEAM / MINIMUM STEAM NEEDED	1	3	1	3
ELECTRICAL ENERGY (KWH) (94 % EFFICIENCY)	1,74	2,48	1,74	2,67
TOTAL	1,1	1,7	1,18	1,95
OUTPUT				
CO	0,77	0,77	0,64	0,64
H_2	0,082	0,082	0,092	0,092
CO_2	0,24	0,24	0,44	0,44
H_2O	0	0,60	0	0,76
TOTAL	1,09	1,69	1,17	1,93
MASS EFFICIENCY (K MOL. (H_2 + CO) PRODUCED PER KILO DRYWOOD)	0,0691	0,0691	0,0691	0,0691

Figure 9

WOOD STEAM GASIFICATION WITH OXYGEN COMBUSTION

MASS BALANCE PER KG OF DRYWOOD
OF 20 % MOISTURE

- - -

	H_2/CO = 1,5	H_2/CO = 2
INPUT		
- DRYWOOD OF 20 % MOISTURE		
. WOOD	0,8	0,8
. WATER	0,2	0,2
WATER / MINIMUM STEAM NEEDED	1,79	3,63
OXYGEN	0,37	0,36
TOTAL	1,37	1,36
OUTPUT		
CO	0,511	0,43
H_2	0,054	0,062
CO_2	0,65	0,77
H_2O	0,15	0,09
ASHES	0,005	0,005
TOTAL	1,37	1,36
MASS EFFICIENCY (K MOL H_2 + CO PRODUCED PER KILO OF DRYWOOD)	0,0459	0,0464

Figure 10

GASIFIER WOOD BURNING CAPACITY

	MARCOUSSIS GASIFIER	PROJECT
(1) PYROLYSIS AND GASIFICATION VOLUME		
HEIGHT	7 METER	11 METER
DIAMETER	0,6 METER	1,3 METER
VOLUME RATIO	1	7,38
(2) COMPUTED WOOD CAPACITY AT 9 BARS (WOOD WITH 30 % MOISTURE)	13,2 T/DAY	97 T/DAY

PROPOSED 20 TONNES PER DAY BIOMASS GASIFICATION PILOT PLANT

Authors : H.T.Wilson, R.Fletcher, R.J.Davies

Contract No. : ESE/P/016/UK

Duration : 24 months 1 January 1982 - 31 December 1983

Total Budget : £1,181,000 CEC contribution : £314,500

Head of Project : Mr. R.M.V.Beith, Research & Development Division

Contractor : Foster Wheeler Power Products Limited

Address : P.O. Box 160, Greater London House,
Hampstead Road, London, NW1 7QN.

Summary

Sized and dried biomass is converted to charcoal at 500°C and essentially atmospheric pressure, by combustion gases from an inert gas generator. Liquid and gaseous pyrolysis products are cooled, the gas being passed to the inert gas generator, drier or flare and the pyroligneous liquor pumped to a slurry mixer. Product charcoal is passed to the mixer, and the resulting charcoal slurry is pumped to the entrained bed gasifier operating at 55 bar pressure. A mixture of oxygen and steam is introduced to initiate and maintain gasification at approximately 1350°C. Steam for gasification is provided by a gas-fired package boiler with superheat facilities. The product gases pass through a shift chamber/quench cooler, then into a scrubbing system where recirculated condensate is used as the coolant. Excess liquor from the scrubber is passed back to the slurry unit. The product gases from the scrubber are analysed and then combusted in the boiler and/or flare/stack facility before discharge to atmosphere. Part of the main product gas from the demonstration plant will be sent through a methanol process route to generate small quantities of methanol for verification and analysis.

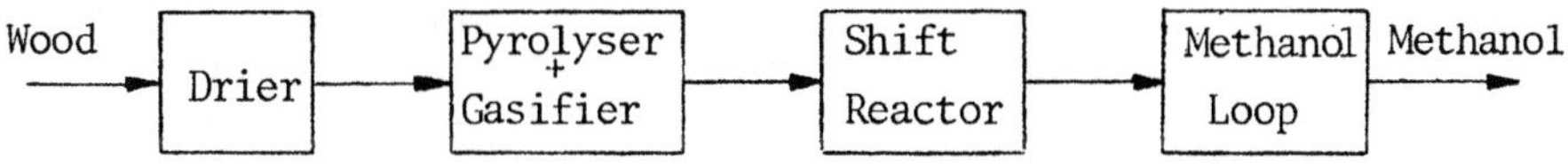

1.0 Introduction

Foster Wheeler Power Products Ltd. (FWPP) have an on-going development programme in the field of pyrolysis/gasification of agricultural, municipal and industrial waste materials. This programme incorporates the construction and operation of a 40 kg/h test facility, partly funded by the EEC. (Contract ESE 106/UK).

As the test facility is still under construction, test data is not yet available to finalise the decision on the most suitable reactor for the 20 t/d test plant now under consideration. Some information will be available early in 1982. This will be used to confirm the selected reactor design.

The present position is that a comprehensive theoretical support study has been undertaken in relation to the performance of biomass gasification systems. In particular, computerised assessments of gasifier-based thermodynamic and kinetic models have been prepared to describe the wide ranging operating conditions that could be experienced on the test facility and other similar gasification plant. The results of this work, plus mechanical design and experimental data derived from the test facility, will be available to confirm the selection of reactor type and ancillary components.

Since submitting our initial proposals to the EEC regarding construction and operation of the 40 kg/h test facility and in addition to acquired practical experience coupled with the development of thermodynamic and kinetic models, an embracing literature review has been undertaken. The results of this study have allowed a more critical systems review to be made and also preliminary selection of the 'optimum' system.

The final choice of reactor system will be guided by the above findings and the data base made available by the members of a proposed operating consortium in order to select the appropriate reactor unit and sub-systems to suit the process requirements.

The collaborators will be selected with a view to combining and providing the maximum related experience and technology into the project. It is anticipated that collaborators will include members of the Foster Wheeler group and companies whose business is plant owning and operating to balance Foster Wheeler's interest which is primarily that of plant design and management of plant supply contracts.

The proposed plant will be designed to investigate the effects on performance of a number of parameters such as feed stock type and properties, operating temperatures and pressure, quantities and relative proportions of the reacting gases, etc.

A provisional study and work schedule will be drawn up in the form of a time/work sequence analysis. A summary of this is presented in Figure 7.

2.0 Description of the Project

The anticipated layout and operation of the pilot plant is as follows:

The as-received biomass for the basic test programme, if in the form of tree trunks, is moved from local storage by means of a front end loader, on to a conveyor which transfers it to a chipper. The biomass, as wood chips, is then fed to a rotary drier which dries the feed from as-received to $< 10\%$ by weight of moisture. This drier unit can operate independently for start-up by firing natural gas (CH_4) - normally it will burn part of the pyrolysis off gas which, in the pilot plant is being directed to flare. A commercial plant would utilise this fuel gas elsewhere in order to achieve energy economy. Drier exhaust gases pass to atmosphere via a gas clean-up system.

The biomass emerging from the drier is passed to a char formation stage via a weigh feeder. The char formation step will be based on the Cross-Flow Pyrolyser, see Figure 1. The feed passes in plug flow through the unit where its lignin structure is destroyed. Based on 463 kg/h oven-dried wood, 128 kg/h of char is produced along with 252 kg/h of pyroligneous liquor and 151 kg/h of gas. Use of a pyrolyser permits a wide range of biomass to be processed, as the system is relatively insensitive to ranges of feed sizes.

Part of the gas fraction produced in the pyrolysis step is recirculated to the inert gas generator where it is combusted to raise the recirculating gas stream temperature to about 500°C. The pyroligneous acids and residual moisture are condensed in a high efficiency venturi scrubber coupled with a heat exchanger. Product char is pulverised and mixed with the pyroligneous acid stream from the scrubber, and make-up water is added as necessary. The resulting slurry is pumped at a high pressure to a partial oxidation reactor. In this way, pyrolysis is undertaken at essentially atmospheric pressure, thus avoiding the difficulties encountered in feeding fibrous material into a high pressure gasifier environment without the escape of toxic or combustible gases. The partial oxidation step will operate at the pressures which could be used on a commercial plant where the low pressure methanol process is being employed, e.g. ICI process (50 - 100 bar).

At the partial oxidation reactor, the slurry temperature is elevated using waste heat. The slurry is injected into the reactor along with superheated steam, and oxygen taken from bulk liquid oxygen storage via a steam vaporiser. The steam is generated by firing either imported natural gas or recovered synthesis gas. For a feed of 'wet' char (quenched) plus pyroligneous liquor of 387 kg/h, approximately 310 kg/h of oxygen will be required to produce the required synthesis gas.

Product gas from the partial oxidation step passes through a refractory-lined cyclone prior to entering a shift chamber where it is contacted with saturated steam. This produces 1193 kg/h of gas having the theoretical volumetric composition : 7.5% CO, 18.7% CO_2, 16.8% H_2, 56.9% H_2O, 0.1% CH_4. This gas is quenched and cooled prior to being discharged from the system, Table 2. The intention is that part of this product gas stream (at pressure) will be passed into a small scale 'once through' methanol production process, to simulate actual methanol yield. In the pilot plant, the medium calorific value synthesis gas will either be flared, used to offset prime fuel imports to the main steam raising plant and/or used for pre-drying the feed to the char producing step of the system.

The ash product, which will contain generally $<5\%$ carbon, can be recycled to the slurry mixing unit, depending on its nature.

Ash in biomass is normally $<1\%$ thus in the partial oxidation step it will only be removed on a discontinuous basis - if necessary slag flux agents will be added.

For start-up purposes, fuel gas and/or oxygen will be fed to the partial oxidation step of the process to thermally condition the plant for process operation.

It should be noted that the proposed operation programme for the partial oxidation step will have an input based upon the results of the experimental work planned for the 40 kg/h scale unit now under construction (EEC Contract ESE 106/UK). The product gas stream leaving the system will be measured and analysed, it will also be assessed by means of a proprietary methanol loop programme available to FWPP and data from equilibrium studies which are being actively progressed.

Excess condensate removed in the scrubber, at the export point, will be stored for recirculation as make-up liquor for the slurry stage - this feature will result in minimum pollution problems from export streams.

The process flow diagram for the proposed pilot plant is shown in Figure 2. A summary heat and mass balance is presented in Figures 3 and 4, and Table 1.

3.0 System Considerations

The proposed system exhibits a number of potential advantages which are summarised below.

1. The cross-flow pyrolyser is very flexible with regard to the nature of the feed material. A wide range of biomass feedstocks can be handled, provided the moisture content is not excessive (i.e. $< 30\%$) and the form is suitable for efficient cross-flow reactor operation.

2. The pyrolysis step is non-polluting since the effluent streams are utilised in the process.

3. Less than one per cent of the energy in the feed is required for the additional power requirements of the pyrolysis stage, i.e. for blowers, pumps, auger, - etc.

4. The heat removed in the venturi scrubber in the pyrolysis stage is recovered in the slurry mixer; operation of the scrubber and mixer is arranged to ensure a small net import of make-up water.

5. The pyrolysis operation can be regarded as a feed preparation stage leading to the formation of char. As this is carried out at essentially atmospheric pressure, solids feeding difficulties are minimised. The final char slurry feed is subsequently pressurised to about 55 bar by means of the slurry pump. For a commercial operation, higher pressures could be employed, although the economic benefits of such a decision have not yet been quantified.

6. The Entrained Bed Gasifier operates at a high temperature ($1360^{o}C$) which results in a syngas product that contains low levels of methane and residual hydrocarbons.

7. The low methane yield suggests that a methane reformer might not be required in a commercial plant.

8. Operation at high pressure leads to a small, compact and close-coupled system. Furthermore, a compression stage before the methanol loop is not required thus minimising the compression energy penalty.

9. The proposed plant is NOT energy optimised, but running costs would be offset by use of the two product gas streams, e.g. up to 40% of product gas energy could be utilised for feed drying, 20% for steam raising,etc. All product streams will be instrumented for data assessment leading to an optimised design of a commercial scale plant.

10. The plant is highly amenable to scale-up, with the pyrolysis stage being duplicated if necessary.

11. During normal operation, the pyrolyser uses its own product gas to supply the heat for char production.

12. The solid residue from the gasifier can be removed, on a batch basis, either in slag form or as ash. The quantity is small and the by-product is inert.

13. The plant's oxygen demand is reduced and reaction efficiency enhanced by the use of low grade heat in the char slurry stream.

14. An oxygen plus steam gasifying medium produces a higher quality synthesis gas, Fig. 5 (1).

15. The efficiency of the plant, based on oven-dried feed to theoretical methanol yield, translates to a mass efficiency, defined as kmol of $H_2 + CO$ in the syngas per kilogramme of feed (20% moisture/80% dry matter) of 2.66%. The plant efficiency, i.e. enthalpy content of equivalent methanol production per unit of enthalpy input to the pilot plant, has been estimated at about 40%. Gasification efficiency, as defined below, is estimated at 95%. A limited review of the literature, and theoretical studies by FWPP, has exposed the inadequacies in quoted system efficiencies. There is an urgent need to totally define the efficiency accurately, accounting for ALL input and output streams that will be present in a commercial plant, so that any conclusions drawn from quoted figures are not distorted. This will be most important when assessing the relative merits of differing process routes.

The following definitions have been used for the pilot plant :-

$$\text{Gasification Efficiency} = \frac{\text{Enthalpy of product syngas}}{\text{Enthalpy of feedstock}}$$

$$\text{Plant Efficiency} = \frac{\text{Enthalpy of product methanol}}{\text{Enthalpy of process feeds}}$$

A plant configuration suitable for overall optimisation, as distinct from process validation, has also been produced. A realistic evaluation of this plant's commercial feasibility, which must include anticipated capital cost and utilisation factor as well as simple conversion efficiencies, forms an essential part of FWPP's Biomass Business Development Plan.

4.0 Financial Considerations

A preliminary cost analysis for a 200 tonnes/day biomass to methanol plant is presented in Figure 6. The assessment is presented on a single 'module' basis. Depending on the level of investment, the modules could be multiplied.

To check the sensitivity of capital investment, including feed handling and civil engineering, on the annual operation costs, two cases have been considered : Case 2 capital charges are approximately twice those of Case 1.

The critical feature of the analysis is the payback period which relates the methanol sales revenue to the net annual operating costs. It may be noted that the payback periods are 7.5 and 12.5 years respectively for the two cases under review, and therefore plants of this size may be only marginally economical at the present time.

A key to project viability is the methanol credit figure, though this is obviously a simplification : for strategic purposes, a particular Government may subsidise the project, either by reducing feed charges or fixing the selling price of the methanol. Increases in methanol credit would significantly reduce the payback period for a commercial plant.

5.0 References

1. " General thermodynamic models for gasifier systems based on chemical equilibria in a C/H/O mixture in the presence/absence of solid carbon" J. R. Gibbins.
(Confidential unpublished Foster Wheeler technical notes).

TABLE 1

Mass and Enthalpy Balance Data

Nature of Stream	Mass Flowrate (kg/h)	Temperature (°C)	Total Enthalpy (MJ/h)
Dried Feed (10% H_2O)	463.0	25	7685.8
Air to Inert Gas Generator	68.7	25	-
Char Product	128.2	500	4373.9
Pyrolysis Gas Export	151.3	40	324.4
Cooling Water from Scrubber	-	-	1294.4
Effluent Pyrolysis Liquor	252.2	40	1308.8
Char Slurry	387.2	250	5961.9
Oxygen to Gasifier	310.0	150	36.0
Ash from Gasifier	4.6	924	4.3
Steam to Gasifier + Shift Reactor	500.0	270	1342.6
Product Syngas	1,192.6	924	7286.2

TABLE 2

Typical gas analysis

WET PRODUCT GAS (SHIFTED)

TEMPERATURE = 924°C
PRESSURE = 55 BAR
COMPOSITION (% VOL): –

H_2	16.9
CO	7.5
CO_2	18.7
CH_4	0.08
H_2O	56.9

SCRUBBED GAS

PRESSURE = 55 BAR
COMPOSITION (% VOL): –

H_2	67.3
CO	29.9
CO_2	2.5
CH_4	0.5

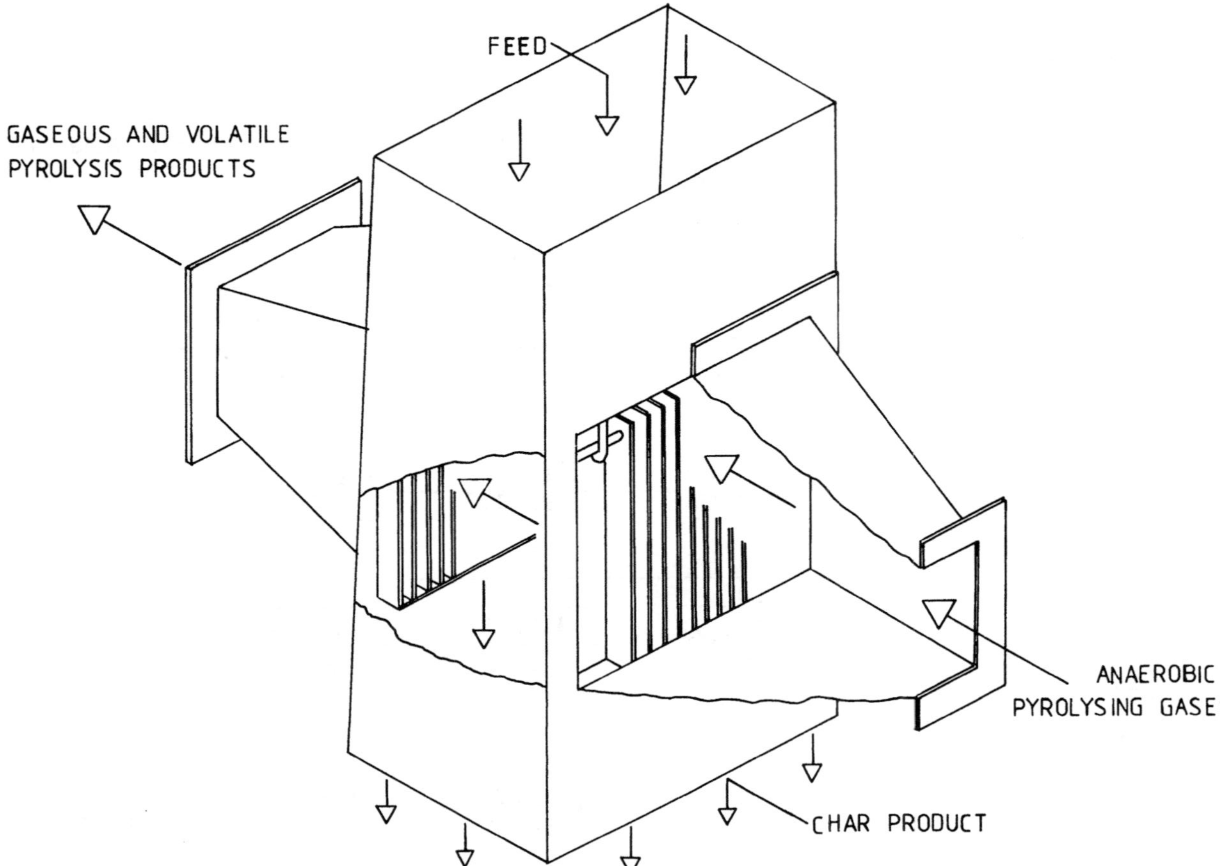

SCHEMATIC CROSS-FLOW PYROLYSIS REACTOR LAYOUT. FIG. 1

FLOWSHEET FOR 20 TONNES/DAY BIOMASS TO METHANOL PLANT

FIG. 2

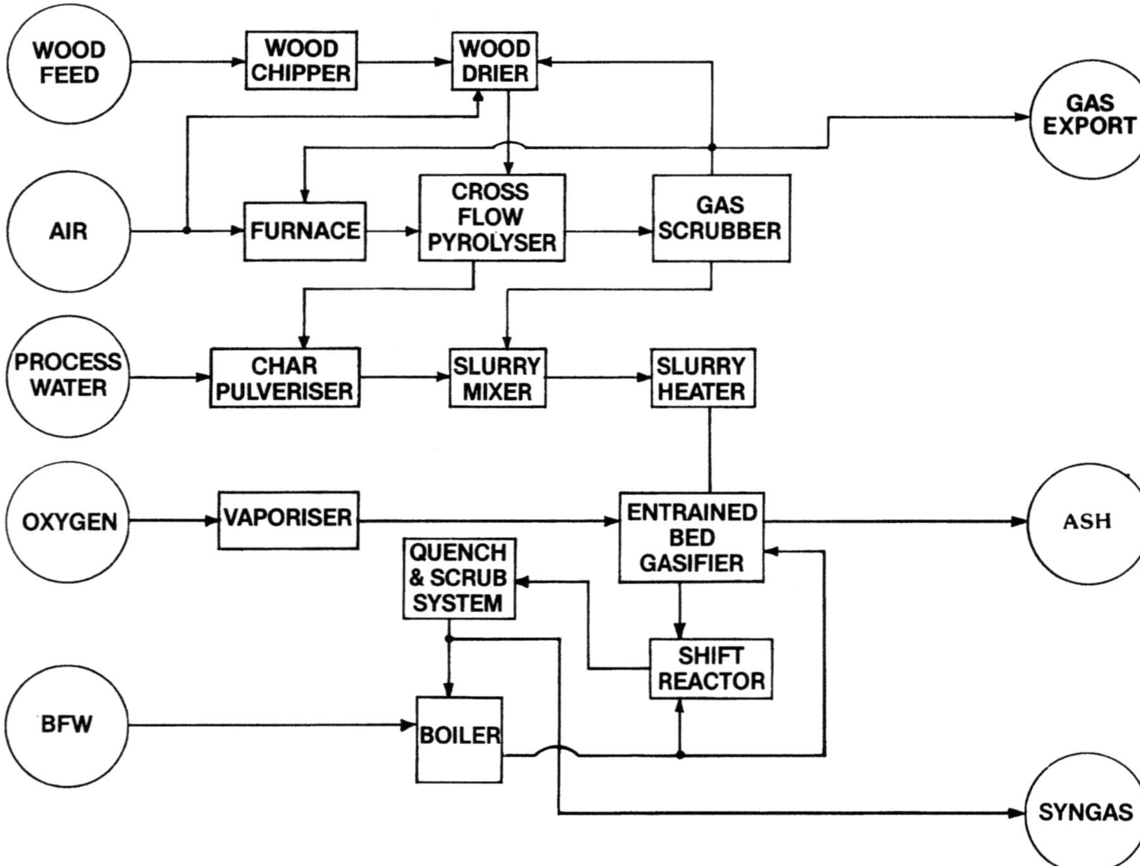

MASS BALANCE FOR 20 t/d BIOMASS TO METHANOL PLANT

FIG. 3

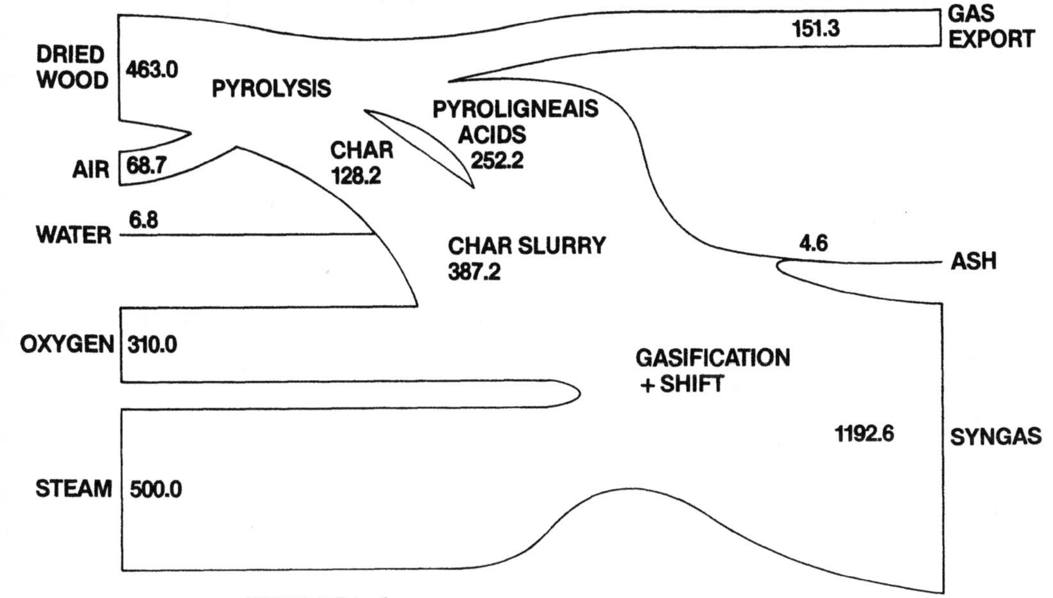

UNITS OF kg/h

FIG. 4

HEAT BALANCE FOR 20 t/d BIOMASS TO METHANOL PLANT

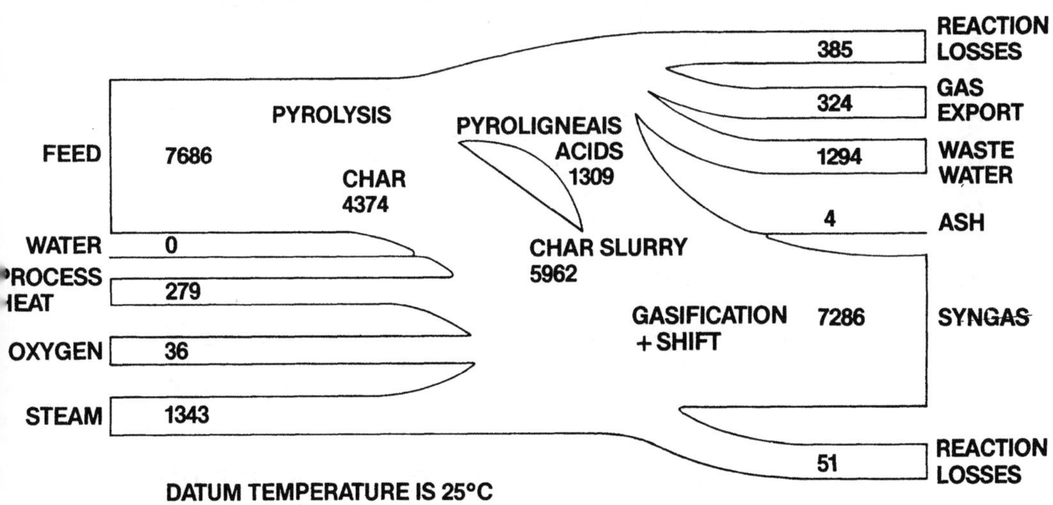

DATUM TEMPERATURE IS 25°C
UNITS OF MJ/h

FIG. 5

TRENDS RELATING TO THEORETICAL METHANOL PRODUCTION FOR VARIOUS STEAM AND OXYGEN FLOWRATES

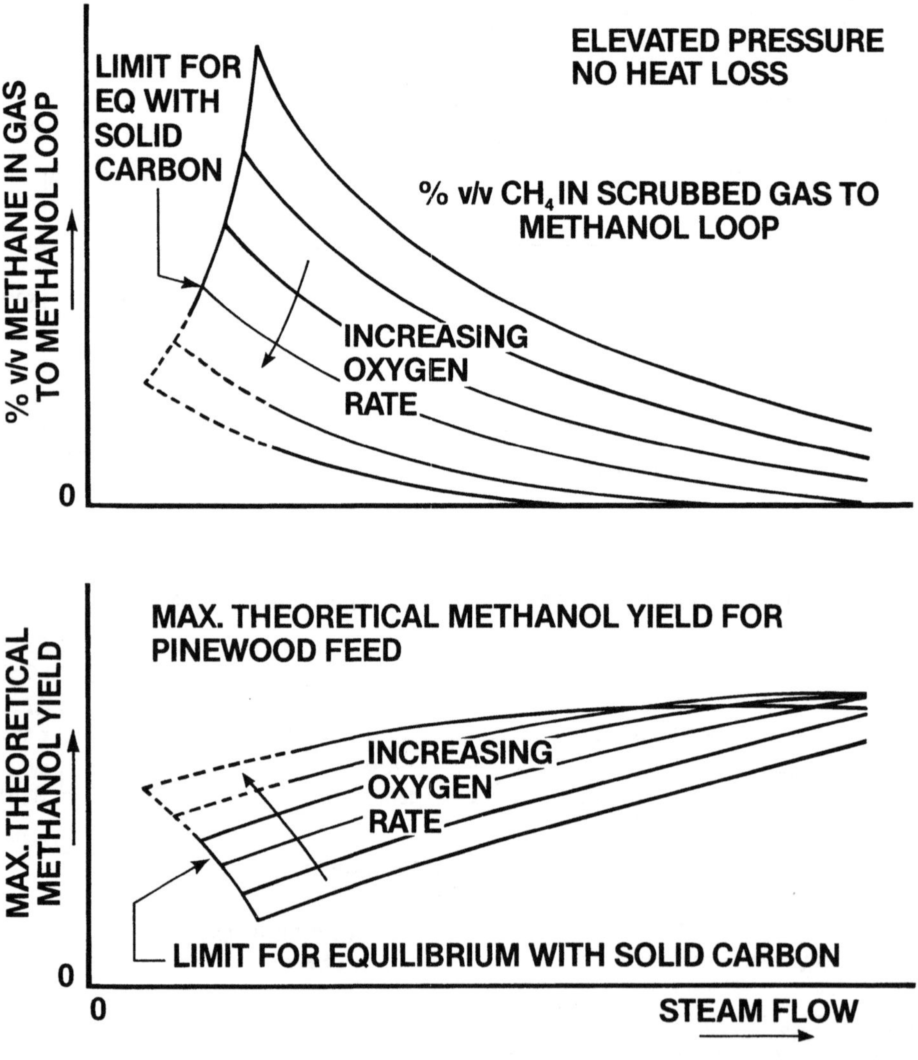

---- $H_2/2CO$ RATIO <1 IN THIS REGION AND RESULTS INVALID. LINES SHOWN ONLY TO DEFINE LIMIT FOR SOLID CARBON EQUILIBRIUM.

COST DATA FOR 200 t/d
BIOMASS TO METHANOL PLANT

FIG. 6

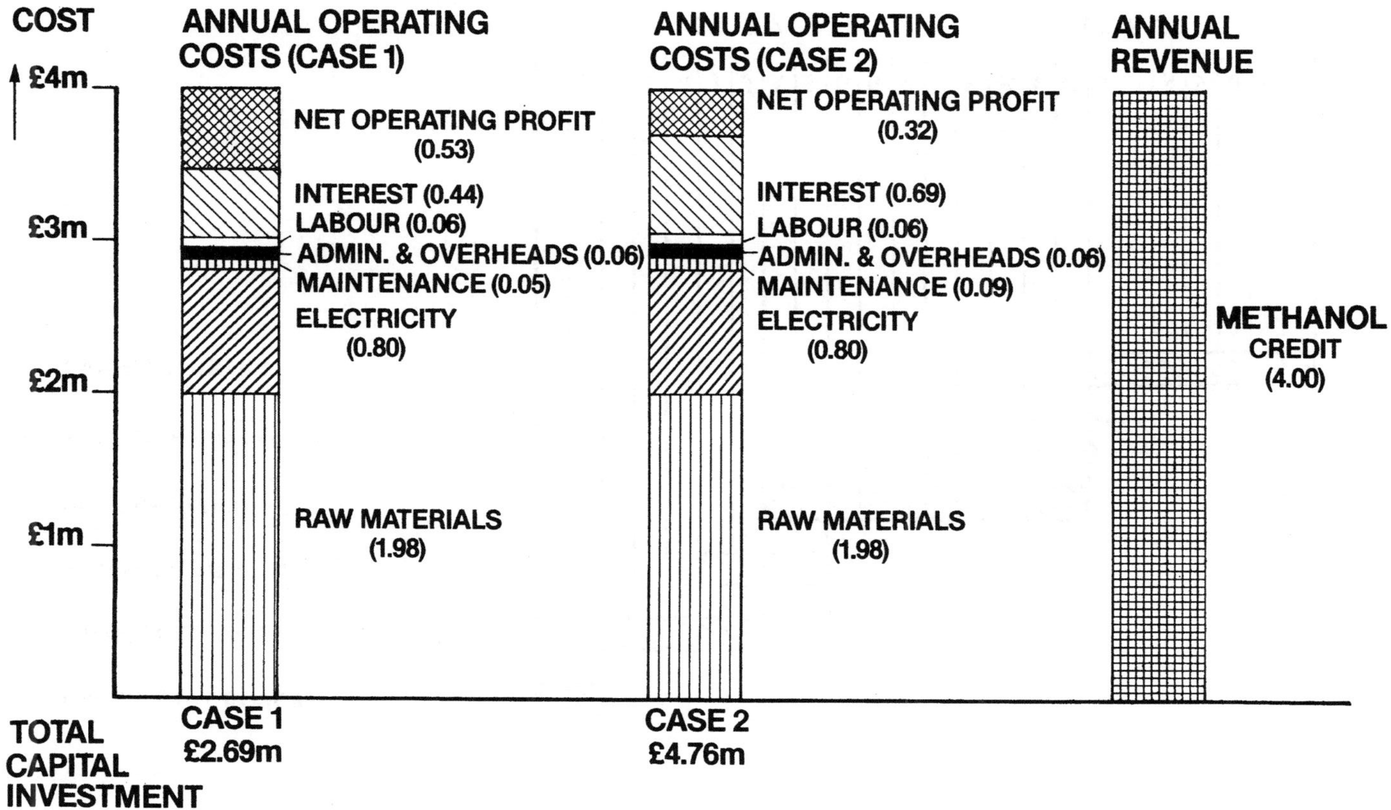

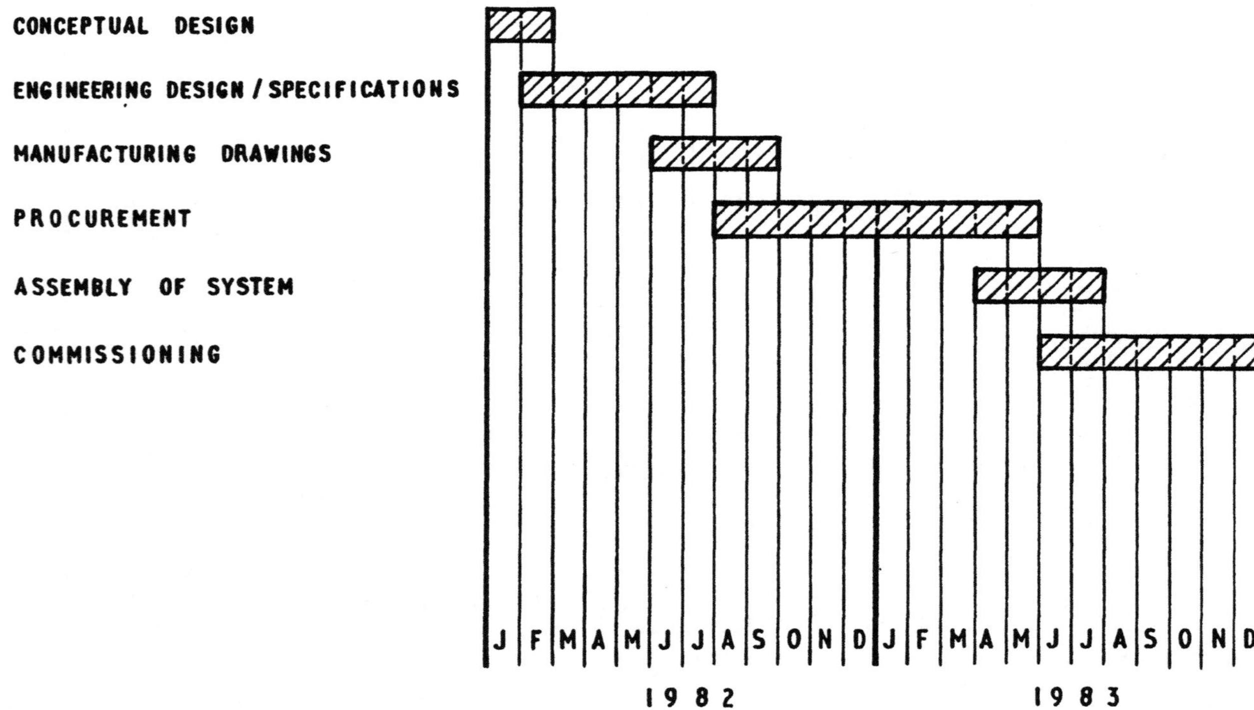

F.W.P.P. LONDON.

PROJECT TIMETABLE

FIG. 7

SYNTHESIS GAS OBTAINED FROM BIOMASS

Authors : R. JANESCH, FRITZ WERNER /
DOTT. M. DE SANTI, REGIONE TOSCANA

Contract number: ESE/P/017/I

Duration : 24 months

Total budget : 1, 560,000 ECU

Head of project: R. Janesch, Messrs. FRITZ WERNER

Contractor : Consortium "Biomasse Maremma"

Address : Regione Toscana, Firenze

Summary

The aim of this paper is to show how synthesis gas is extracted from biomass while ensuring minimal strain on the environment and maximal utilization of biomass energy.

In our previous method combustion in the reactor occurred at 1,660 K and gasification at 1,300 K, whereas in this process the gasification phase takes place at 1,000 K with a combustion temperature of 1,200 K. Unwanted constituents, mainly tar and methane,which are formed at the low gasification temperature are split up in a cracking chamber.

A thermodynamic evaluation of the combustion, gasification and cracking processes in the reactor and the heat balances for the whole plant show that when the residual moisture of the biomass is 20 % and the oxygen fed into the cracking chamber is 8.5 % of the biomass throughput, then the proportion of the gas extracted will be approximately 58 %.

The gasification reactions have been calculated with the aid of the EDP programme at the ICT Institute for Fuels and Explosives in Pfinztal. The basis for the thermodynamic calculations was 100 % initial moisture, 20 % residual moisture, and the throughput at 20 % residual moisture was taken as 1,000 kg biomass per hour. The lower calorific value is 13,850 KJ/kg.

The gas produced by the pilot plant will be used in a pumping station and to heat a greenhouse. Other uses can be identified in the area if the first mentioned are not able to accept the complete gas quantity.

1. 1 Introduction

There are several ways of converting biomass into useful energy, for example, aerated gasification, steam production by combustion, gas production by chemical decomposition (putrefaction) etc.

This investigation describes the production of synthesis gas by gasifying biomass to the exclusion of air, and shows the results of the thermodynamic evaluation of such a gasification plant. It should be noted that the calculations connected with the gasification reaction were carried out with the aid of a special EDP programme at the Fraunhofer Institute for Fuels and Explosives Chemistry. A detailed description of the calculating procedure can be dispensed with here since the computing programme has been tested by the German State Quality Control Association, and therefore accepted by the German authorities.

These calculations were based on a substitute mass whose composition is identical to that of beechwood.

The gasification temperature and the addition of oxygen and water vapour were varied during calculations so that we would be able to determine the optimum operating conditions of the plant and the properties of the synthesis gas.

By choosing a correspondingly large heat exchange area between the combustion chamber and the gasification chamber we were able to reduce the combustion and gasification temperatures to a level where all the reactor components are guaranteed to function with adequate safety over a fairly long period of time. The unwanted components (mainly tar and methane) formed at the relatively low gasification temperature are split up in a special cracking chamber.

The plant's energy and mass balances, which will be shown during the course of the paper, take account of the quantity of gas or energy required to produce electrical energy for the plant's electric drive units. However, the technological details of the plant's individual sub-processes are not the subject of this investigation.

The present proposal is in reply to an EEC tender to promote the realisation of a pilot plant for the production of liquid and gaseous fuels from renewable biomass using advanced technology.

The raw material which is used to feed the proposed plant consists of a wide variety of biomass such as coppice wood, press cake from wet fractionation of grasses, olive rapes, lignite , algal biomass etc.

European State and private agencies intend to use oil from biomass as a substitute for fuel, and within this framework, Regione Toscana is promoting the realisation of a pilot plant for the production of syngas for methanol from the above listed materials in the Maremma district (see map).

As for forestry and agriculture, Maremma district represents a situation which is typical of several other areas in northern and southern Italy. There are also marginal soils present which can be prepared for cultivation through energy plantations.

Finally, Maremma district has been included in the European Community Programme concerning the forestation of mediterranean areas. The gasification process proposed here has been studied with the aim of achieving maximum operational flexibility (as for raw materials) and verifying innovative technologies.

1. 1. 1 Project organization and participants

An international consortium has been formed. It is promoted by the

Regione Toscana and consists of Agip Nucleare, CNEN and the Ansaldo Group for the Italian part, and FRITZ WERNER Industrie-Ausrüstungen GmbH for the German part.

1. 1. 2 Financial aspects of the proposal

The total costs of the project for engineering, construction, installation, operation, testing and consumption material, are estimated at ECU = 1,560,000. If the commission accepts the project with a participation of 600,000 ECU the total amount will be shared as follows:

C.E.C. = 600,000 ECU
FRITZ WERNER = 320,000 ECU
Italian partners = 640,000 ECU

The partners of the consortium have given a written commitment to "Regione Toscana" to support the above mentioned costs.

1. 2. Description of the process

The first part of the process consists of the raw biomass being machined in a chipper or grinder in order to obtain the required size, which is approximately 3 x 15 x 15 mm. It is then dried in a special drying plant. The heat energy necessary for this operation is supplied by the flue gases produced in the reactor's combustion chamber and by the heated air leaving the gas cooler. The hot gases and the exhaust vapour leave the dryer at a temperature of about 100° C via a cyclone system. The heat from these hot gases can be exploited by installing a heat exchanger downstream from the dryer, for example for warm water treatment. After a certain retention time in the dryer the dried fuel is transferred outwards. The residual moisture content is 20 %.

The dried biomass is buffered in a silo in case of possible in-stationary operation. Thereafter the biomass is fed to the fluidized-bed reactor by means of a conveying system. Before gaining access to the reactor the biomass is separated into two parts: one to undergo gasification and the other to undergo combustion.

The reactor is divided into a combustion chamber and a gasification chamber. The ring-shaped gasification chamber is built around the combustion chamber in such a way that the hot gases leaving the combustion chamber are led past the outer wall of the gasification chamber before leaving the reactor.

It is also possible to generate the gasification heat with the synthesis gas produced by using an appropriate burner system and thereby gasify practically all the biomass. One advantage of this method is that no residue whatever is formed in the combustion chamber and therefore the often problematical task of removing ash is obviated. Only when starting the plant must an auxiliary gas be held in reserve for the burner unit, for example a small wood gasifier, or buthane in cylinders.

The temperature in the reactor's combustion chamber is 1,200 K and the biomass is gasified at 1,000 K. The maximum temperature of the heat exchanging wall located between the combustion chamber and the gasification chamber is 830° C.

Because of the relatively low gasification temperature the gas produced contains tar and methane, and these are unwanted constituents when using this gas. This is especially the case with the production of methanol and the drive of internal combustion engines.

This is why the gas is cracked in a fluidized-bed cracking chamber built into the reactor by adding 8.5 % oxygen. The temperature inside the cracking chamber is 1,350 K. It is necessary to provide the cracking chamber with a lining, made of ceramic, for example, so as not to subject the chamber wall to thermal overload.

The gas leaves the reactor at 1,350 K via a cooled cyclone, and is then cooled down to 310 K in a heavy-duty cooler and fed into the gas store via a fine-mesh filter.

The heated cooling air, which leaves the gas cooler at a temperature of around 800 K is required primarily for the combustion chamber of the reactor. The rest is fed in the biomass dryer.

1. 3 Results

The combustion air being 800 K and the gasifying temperature 1,000 K 265 kg of biomass has to be burnt per hour to gasify 735 kg of biomass per hour.

The oxygen added during cracking is 8.5 % relative to a biomass throughput of 1,000 kg/h, and the temperature of the oxygen is about 800 K on feeding.

The quantity of water-free synthesis gas produced is 578 kg/h.

The gas composition shown in the following table gives information about the quality of the synthesis gas:

components	proportions (Vol. %)
H_2	43.1 %
CO	43.7 %
CO_2	4.5 %
N_2	0.04 %
H_2O	8.66 %

It should be noted that the water proportion of 8.66 % is separated in the gas cooler.

The temperature of flue gas in the combustion chamber is about 1,200 K.

The mass efficiency is expressed by the number of kilomoles (kmol) of hydrogen + carbon monoxide produced per kg of dried biomass with a 20 % moisture content. This mass efficiency is 0.03 kmol/kg.

The net input of oxygen needed for the cracking process is about 85 kg/h.

The power consumption of all the plant's auxiliary units is approximately 230 KW h; this value, however, has already been included in the heat balance and the mass-flow balance and it gives no information on the performance of the plant.

The maximum temperature of the heat-exchanging wall located between the combustion chamber and the gasifying chamber of the reactor was 830° C and was calculated on the basis of the reactor operating under full load conditions.

1. 4 Analysis of results and comments

Looking at the composition of the synthesis gas produced the first thing we notice is the relatively high proportion of carbon monoxide and hydrogen. The water is present in the synthesis gas only during the hot stage and separated in the gas cooler; so, the proportion of water is negligible.

As already mentioned , tar and methane which have formed at low gasifying temperature have been completely cracked in a subsequent cracking process. Finally, the proportion of nitrogen as a result of gasification to the exclusion of air is so small (0.04 %) that it may be considered as negligible, too.

The proportion of carbon dioxide (4.5 %) remains within acceptable limits; it can be reduced, if necessary, in a special washing process. So, the extraction of methanol from this synthesis gas may be carried out without any major problems.

Considering that the maximum temperature of the material in the reactor is 830° C it should be noted that a sufficient service life of the reactor components exposed to high temperatures can be ensured; this can be achieved by choosing a highly heat-resistant material which remains sufficient stable under reducing conditions as is the case in the gasifying chamber and under oxidising conditions as is the case in the combustion chamber. Such a material might be, for example, Incoloy 800. This material is resistant to both oxidation and reduction at high temperatures and even at 900° C it has a mechanical stability under load of 4 N/mm^2 during an assumed service life of 5 years. In this respect it should be considered that the internal reactor components practically undergo no mechanical stress. Thus, the stability of these components may be considered as guaranteed.

2. Conclusions

The analysis of the calculation results showed that the low combustion and gasification temperatures are advantageous with respect to the quantity and also the quality of the synthesis gas produced. In addition to this, the stability of the components exposed to heat is substantially improved.

However, these temperatures cannot be lowered below a minimum limit; otherwise, the proportions of tar, methane and especially of solid carbon would be so high that they might cause problems during the cracking phase. Moreover, substantially more oxygen would be needed to meet the requirements of the cracking process and this would negatively affect the efficiency of the gas produced. The analysis showed that the ideal combustion and gasification temperatures are about 1,200 K and 1,000 K respectively.

The theoretical evaluation example for the gasifying process involves a certain quantity of biomass being burnt in order to produce the heat necessary to gasify the remaining quantity.

As already mentioned, it is possible to burn a part of the synthesis gas extracted in order to gasify the total quantity of biomass.

This process has two merits: it ensures on the one hand better adjustability of the reactor's combustion system and on the other hand only one ash discharge unit is necessary.

Considering the relatively low gasifying temperature the heat exchanging wall between the combustion chamber and the gasifying chamber of the reactor would have to be increased in this particular case since all of the biomass is to be gasified.

The efficiency of the plant and the net quantity of the gas produced are negligible modified; the quality of the gas remains completely unchanged.

Last but not least, two outstanding features of the plant are its high performance and its relatively simple design.

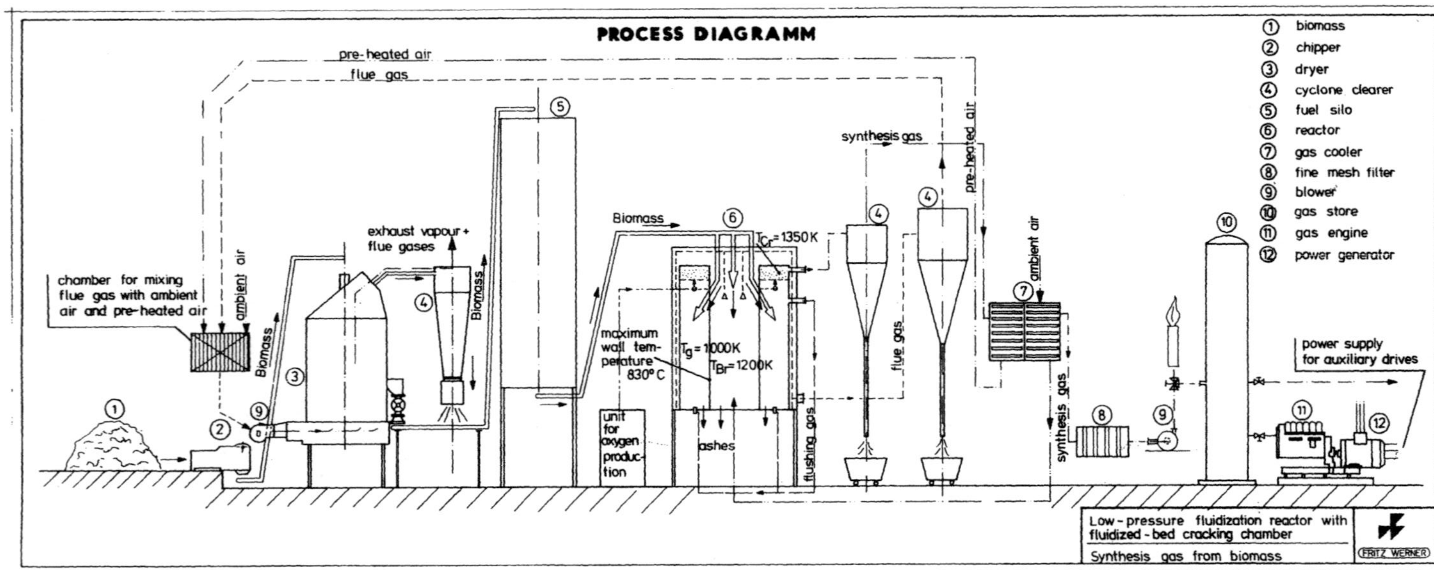
PROCESS DIAGRAMM
pre-heated air
flue gas
chamber for mixing flue gas with ambient air and pre-heated air
ambient air
Biomass
exhaust vapour + flue gases
Biomass
maximum wall temperature 830° C
T_g = 1000K
T_{Br} = 1200K
T_{Cr} = 1350K
unit for oxygen production
ashes
flushing gas
synthesis gas
flue gas
pre-heated air
ambient air
synthesis gas
power supply for auxiliary drives
① biomass
② chipper
③ dryer
④ cyclone clearer
⑤ fuel silo
⑥ reactor
⑦ gas cooler
⑧ fine mesh filter
⑨ blower
⑩ gas store
⑪ gas engine
⑫ power generator
Low-pressure fluidization reactor with fluidized-bed cracking chamber
Synthesis gas from biomass
FRITZ WERNER

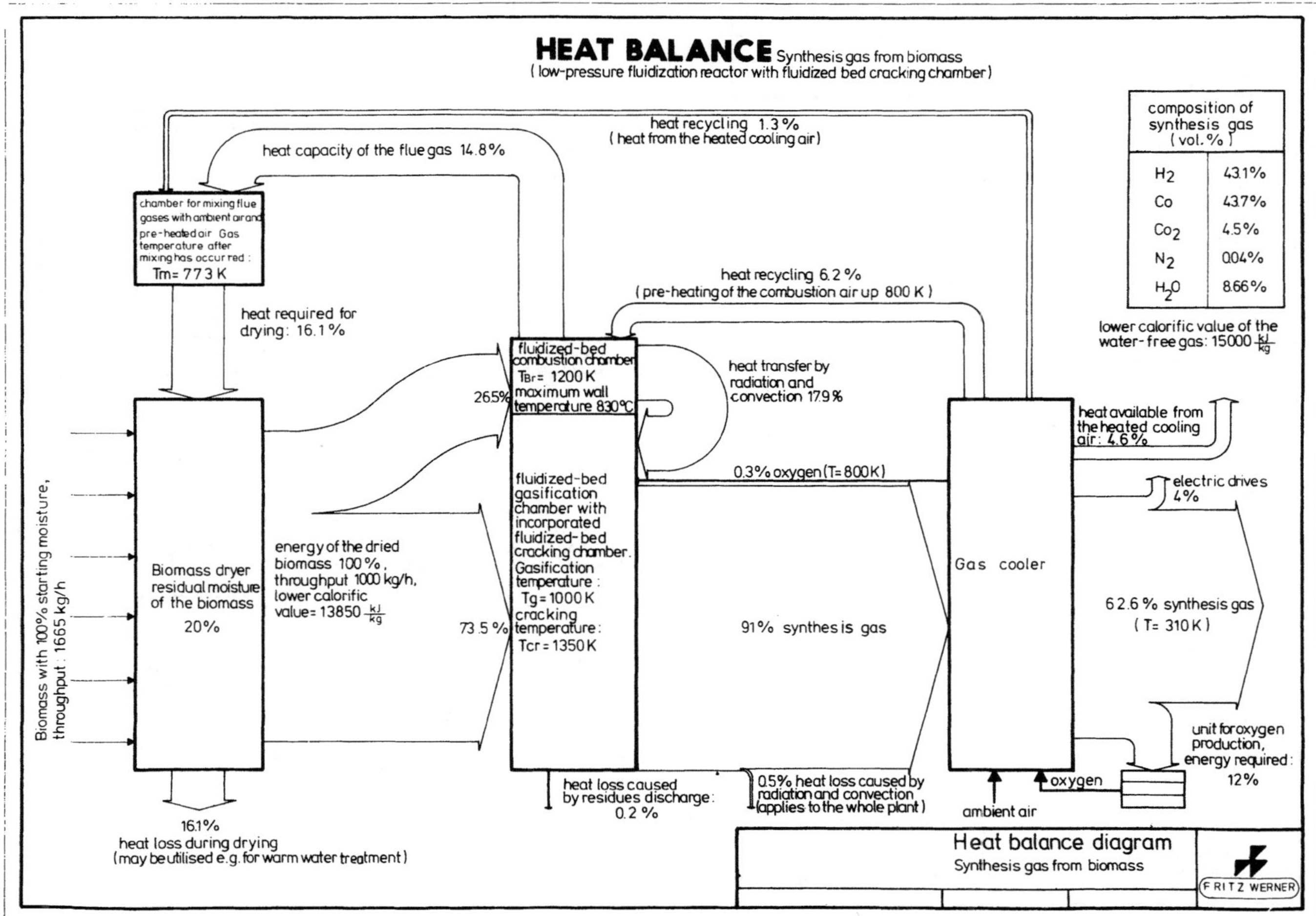
HEAT BALANCE Synthesis gas from biomass
(low-pressure fluidization reactor with fluidized bed cracking chamber)
heat recycling 1.3 %
(heat from the heated cooling air)
heat capacity of the flue gas 14.8%
chamber for mixing flue gases with ambient air and pre-heated air Gas temperature after mixing has occurred :
Tm= 773 K
heat required for drying: 16.1%
heat recycling 6.2 %
(pre-heating of the combustion air up 800 K)
composition of synthesis gas (vol.%)
H2 43.1%
Co 43.7%
Co2 4.5%
N2 0.04%
H2O 8.66%
lower calorific value of the water-free gas: 15000 kJ/kg
fluidized-bed combustion chamber
TBr= 1200 K
maximum wall temperature 830°C
heat transfer by radiation and convection 17.9%
26.5%
heat available from the heated cooling air: 4.6%
0.3% oxygen (T= 800 K)
electric drives 4%
Biomass with 100% starting moisture, throughput : 1665 kg/h
Biomass dryer residual moisture of the biomass 20%
energy of the dried biomass 100%, throughput 1000 kg/h, lower calorific value= 13850 kJ/kg
fluidized-bed gasification chamber with incorporated fluidized-bed cracking chamber.
Gasification temperature :
Tg= 1000 K
cracking temperature:
Tcr= 1350 K
73.5 %
91% synthesis gas
Gas cooler
62.6 % synthesis gas
(T= 310 K)
unit for oxygen production, energy required: 12%
heat loss caused by residues discharge: 0.2 %
0.5% heat loss caused by radiation and convection (applies to the whole plant)
oxygen
ambient air
16.1%
heat loss during drying
(may be utilised e.g. for warm water treatment)
Heat balance diagram
Synthesis gas from biomass
FRITZ WERNER

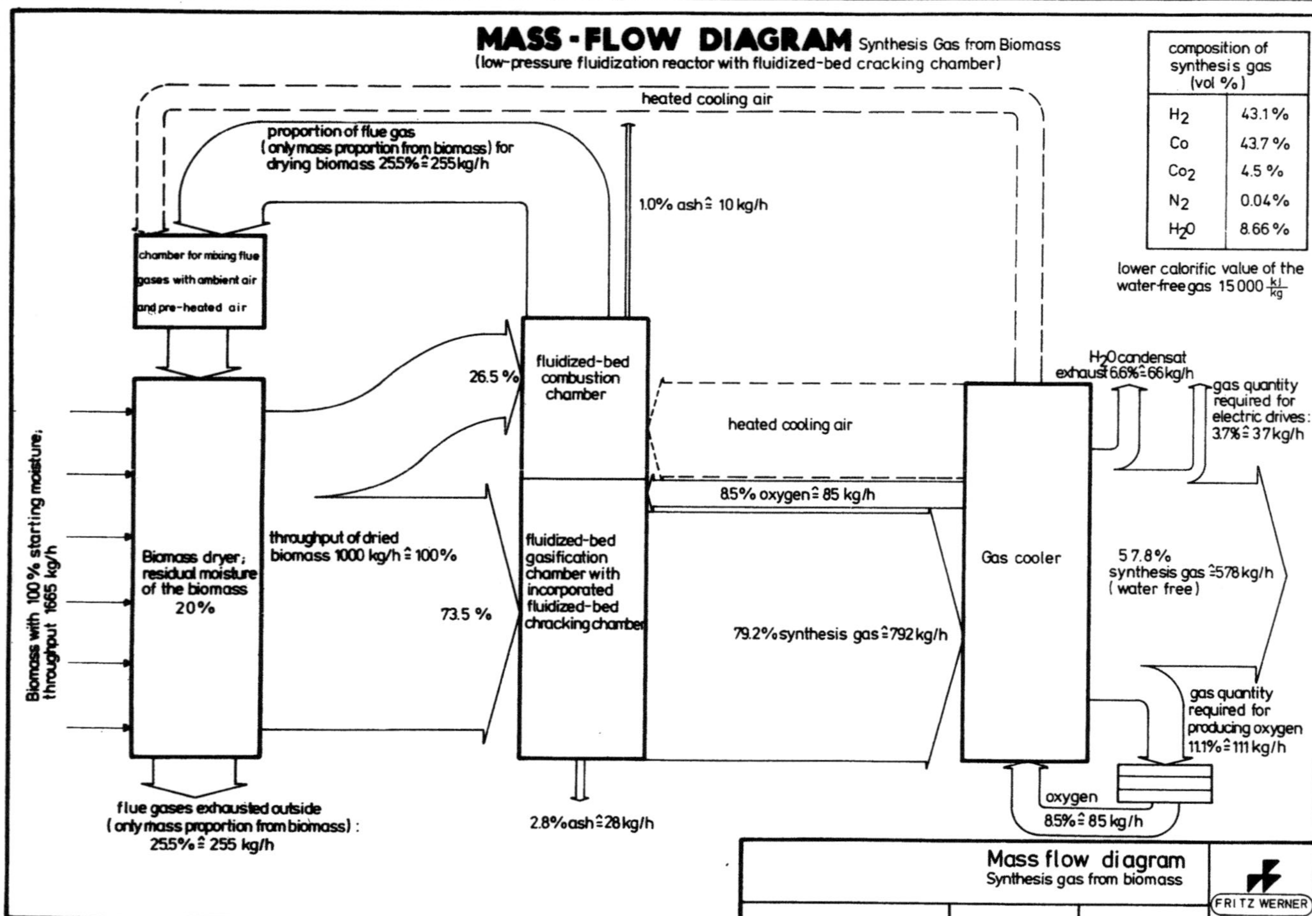

MASS - FLOW DIAGRAM Synthesis Gas from Biomass
(low-pressure fluidization reactor with fluidized-bed cracking chamber)
composition of synthesis gas (vol %)
H2 43.1 %
Co 43.7 %
Co2 4.5 %
N2 0.04 %
H2O 8.66 %
lower calorific value of the water-free gas 15 000 kJ/kg
heated cooling air
proportion of flue gas (only mass proportion from biomass) for drying biomass 25.5% ≙ 255 kg/h
1.0% ash ≙ 10 kg/h
chamber for mixing flue gases with ambient air and pre-heated air
Biomass with 100 % starting moisture; throughput 1665 kg/h
Biomass dryer; residual moisture of the biomass 20%
throughput of dried biomass 1000 kg/h ≙ 100%
26.5 %
73.5 %
fluidized-bed combustion chamber
fluidized-bed gasification chamber with incorporated fluidized-bed chracking chamber
heated cooling air
85% oxygen ≙ 85 kg/h
79.2% synthesis gas ≙ 792 kg/h
Gas cooler
H2O condensat exhaust 6.6% ≙ 66 kg/h
gas quantity required for electric drives: 3.7% ≙ 37 kg/h
57.8% synthesis gas ≙ 578 kg/h (water free)
gas quantity required for producing oxygen 11.1% ≙ 111 kg/h
oxygen 85% ≙ 85 kg/h
flue gases exhausted outside (only mass proportion from biomass): 25.5% ≙ 255 kg/h
2.8% ash ≙ 28 kg/h
Mass flow diagram
Synthesis gas from biomass
FRITZ WERNER

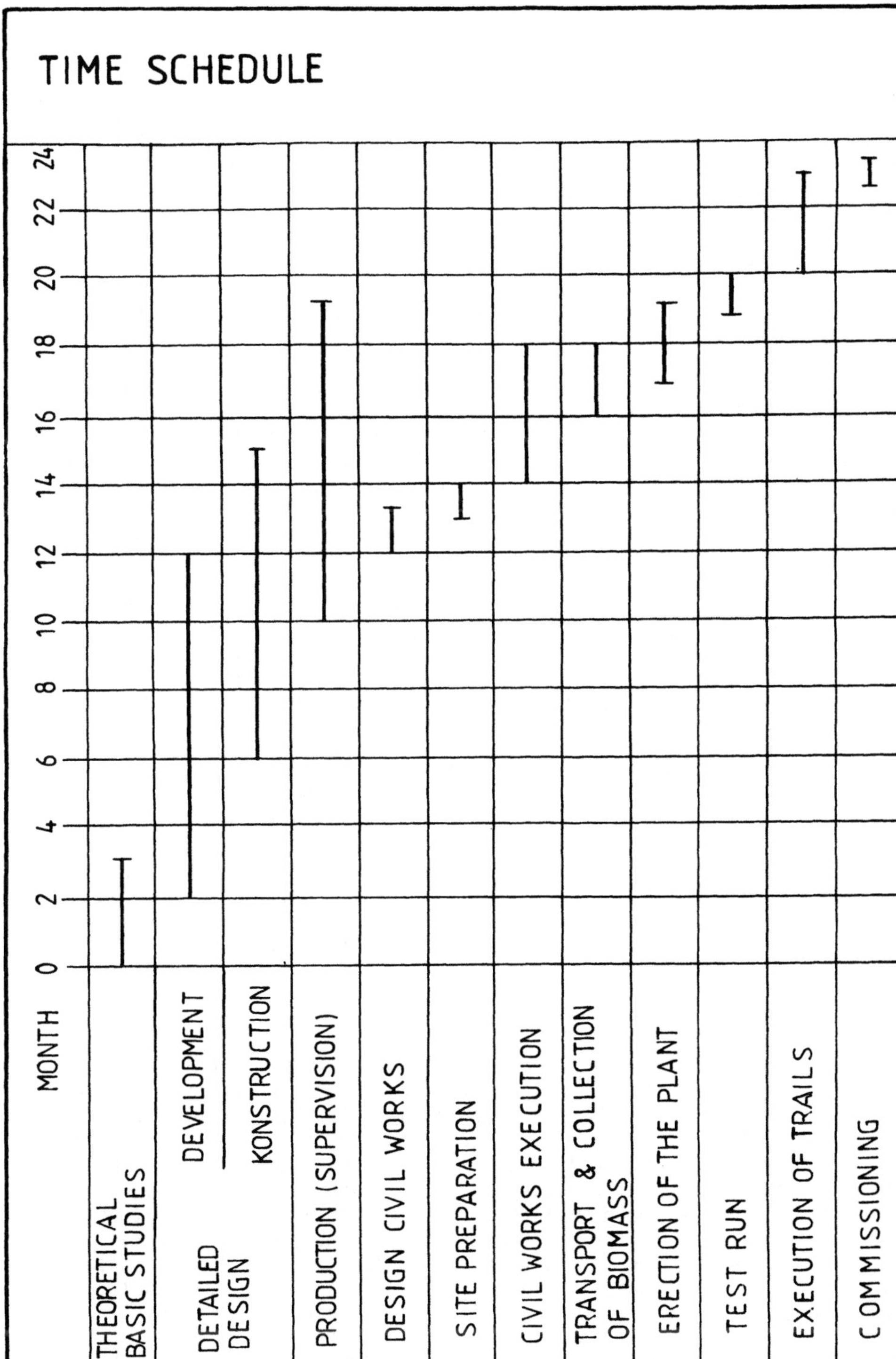
TIME SCHEDULE
MONTH
0
2
4
6
8
10
12
14
16
18
20
22
24
THEORETICAL BASIC STUDIES
DETAILED DESIGN
DEVELOPMENT
KONSTRUCTION
PRODUCTION (SUPERVISION)
DESIGN CIVIL WORKS
SITE PREPARATION
CIVIL WORKS EXECUTION
TRANSPORT & COLLECTION OF BIOMASS
ERECTION OF THE PLANT
TEST RUN
EXECUTION OF TRAILS
COMMISSIONING

3. Materials

One of the features of the pilot plant in the present project is the possibility of using, as mentioned in the foreword, different raw materials, or alternatively different mixtures of these: wood, press cake from wet fractionation of grasses, olive rapes, lignite, algal biomass. Therefore the proposed plant exhibits traits of operational flexibility that enable the siting to take place in different pedoclimatic, agronomic and socioeconomic conditions.

3. 1 Characteristics

The behavioural properties of wood and lignite during gasification are well known to this audience.

The press cake from wet fractionation of grasses is obtained from several different crops by an initial mechanical separation of plant juices, followed by precipitation of proteins which represent a second main by-product of the process (a third being represented by deproteinized sera, useful for both biomass and methane generation). The technology of the wet fractionation of green grasses is well developed in Europe on a commercial level and the properties of the press cake are similar to those of wood (as d.w., according to the extensive studies reported in this proposal) in respect of both caloric content and physico-chemical composition. In fact an additional advantage of press cake over wood is its granulometric composition, which is ideal for gasification on a fluid bed. Moreover, as mentioned above, the anaerobic digestion of deproteinized sera gives rise to methane; the energy contribution of the biogas to the process represents about 25 % of the total energy of the corresponding press cake, and improves the H_2/CO ratio in the syngas.

Olive rapes consist of about 65 % stone and 35 % pulp. Considered together, the two fractions have a caloric content at least the same as that of wood (as d.w.), a physical structure which allows their use on fluid bed without pre-treatment, and an initial moisture of 12 %.

Algal biomass has a composition which allows the recovery of the protein fraction for use as fodder panels or as nutritive support to improve the bioconversions preliminary to gasification of the residues obtained; in the absence of pre-treatment, this raw material can be gasified after the water content has been lowered,this being more amenable than with wood.

3. 2 Availability of resources

The Region of Toscana has a coefficient of 40.6 % wood present on its territory, and about half a million ha are represented by coppice wood.

In the vicinity of the plant siting area (within a radius of less than 15 km) 36,000 ha of wood are actually available, and 34,000 ha have coppice wood on plain land and hills; about 1/3 of the acreage is owned by State Agencies (Municipalities, Provinces etc.). Average productivity of the above mentioned coppice wood is estimated at between 100 and 150,000 metric tons/annum. Therefore, the production obtainable in one year would allow the pilot plant to run, as well as the scaled-up plant in a future perspective. The press cake is readily available, provided a wet fractionation plant is in the vicinity. To obtain sufficient press cake to substitute for the coppice wood in a gasification plant, only 1/5 to 1/3 of the land needed for the coppice wood supply is required. Such an area is largely available in the region specifically involved with the project proposal, and is enough to feed both the pilot and scaled-up plants. When considering press cake, the other advantages deriving from its use must be taken into account: the yield of protein concentrates obtained via wet fractionation of grasses is qualitatively higher (4 t/ha/year at 45-50 % of protein content) than those obtainable with any proteinagineous crop (2-2.5 t/ha/year at 30-35 % protein) such as grain.

The availability of olive rapes in the project area depends on the harvest and storage carried out by industries called "sansifici". In Tuscany the olive production reaches 150,000 t/year, with a production of 40-45,000 t of olive rapes during the year, which are delivered by the factories. Such an availibility supports entirely the needs of the pilot plant, and more than half of the scaled-up plant.

Lignite deposits are available in the Buccinello Cana mines at a distance of about 100 km from the proposed plant. The estimated resources are 20 to 40,000 t, enough to define lignite as an integrator.

The coastal area which could provide algal biomasses extends from a few km to about 100 km from the pilot plant site. In 1976 the primary productivity was extensively studied in the Orbetello lagoon, which gave as an average from June till November of 2.4 to 2.8 g/m³ /day of organic matter; in that area algae are alreday being harvested on a (local) commercial level and supplies for the pilot plant can be organized.

3. 3 Cost of the raw material

The average costs of selected coppice wood, suitable for domestic uses, in the proposed area fall in the range of 5.5 to 6.5 ECU per 100 kg (airdried). However the pilot plant can also be fed with wood of more inferior quality than the one for domestic uses only. Moreover these prices can be lowered in the context of larger commercial harvesting units.

The press cake from wet fractionation of grasses is available on a commercial level in Italy as fodder for ruminants, at prices ranging 3.5 to 10 ECU/100 kg (pelletted and at 12 % moisture) depending upon the producer as well as on the quality of the product. The lowest prices are the maximum one could expect when the press cake is produced within the gasification process, which includes a wet fractionation plant.

Olive rapes are sold at prices ranging 3.5 to 4.5 ECU/100 kg (12 % moisture) at present.

For lignite and algal biomasses, market prices have to be defined largely according to specific requests and/or to extraction separation costs.

4. Site

As said before the pilot plant is to be sited in an area in the Province of Livorno, namely the "val di Cornia" zone which includes the municipalities of Suvereto, Sassetta, Campiglia, Marittima etc. The Authorities expressed their agreement with the initiative during a meeting held in the last days of September.

The surface area is 471.69 square kilometres, corresponding to 2.05 % of the whole regional extension.

From the orographic point of view, the area includes Monte Calvi (646 m. a. s. l.), the highest mountain in the province of Livorno: along its sides the municipalities of Campiglia, S. Vincenzo and Suvereto border. The west sides of the Colline Metallifere chain are also in the area of study as well as Monte Massoncello, the most important massif of the Piombino promontary (286 m a. s. l.).

As far as the hydrography is concerned, particular emphasis must be given to the plain crossed by the Cornia river and its tributaries lying between Monte Massoncello and Monte Calvi. The reclamation of this area has been completed in the past few years and the Cornia river estuary has been moved artificially eastwards. A remarkable amount of underground water is present.

The surface area of the plains is approx. 186 square kilometres while the areas with a gradient inferior to 10 % have an extension of 211 square kilometres.

The hydrogeological assessment is characterized by a good stability and there isn't any landslide phenomenon in the area. The aquifer network includes the whole area of Sassetta, a great part of the Monteverdi Marittima, the highlands from Piombino to Baratti and the hilly zone north of Riotorto.

The low and hilly areas are prevailingly calcareous and sufficiently fertile; a remarkable presence of lime can be found in the areas lying in the plain due to flooding (also of artificial nature); the soil is here rich with mineral salts, fresh, calcareous and extremely fertile.

In the Massera river valley (one of the right tributaries of the Cornia river) a clayey belt makes the soil rather poor and unsuitable for agriculture purposes.

The climate of the whole area is definitly favourable: the Val di Cornia together with the Crosseto plain is the sunniest part of the whole of Toscana. Both rain and hail are very scarce (less than 700-800 mm per year).

The whole area is supplied with an excellent transport network. As far as the railway system is concerned apart from the Genova Roma line, worthy of mention are the single-track electric line Campigliano-Piombino, which is being improved and the link between the station and the harbour of Piombino activated in 1961.

The road network is very efficient in particular in respect of highways, freeways and provincial roads. Besides the Autostrada del Sole and the Aurelia, the Province of Livorno is supplied by the following network:

the "Della principessa" road linking San Vincenzo to Piombino; the double access from the Aurelia to Campiglia Marittima (via Madonna della Fucinaia northwards and via Venturina westwards), the Suvereto-Sasseta-Bocca di Valle road, the Suvereto-Forni road. The linking between the "Della Principessa" and the Aurelia roads is soon to become of provincial importance.

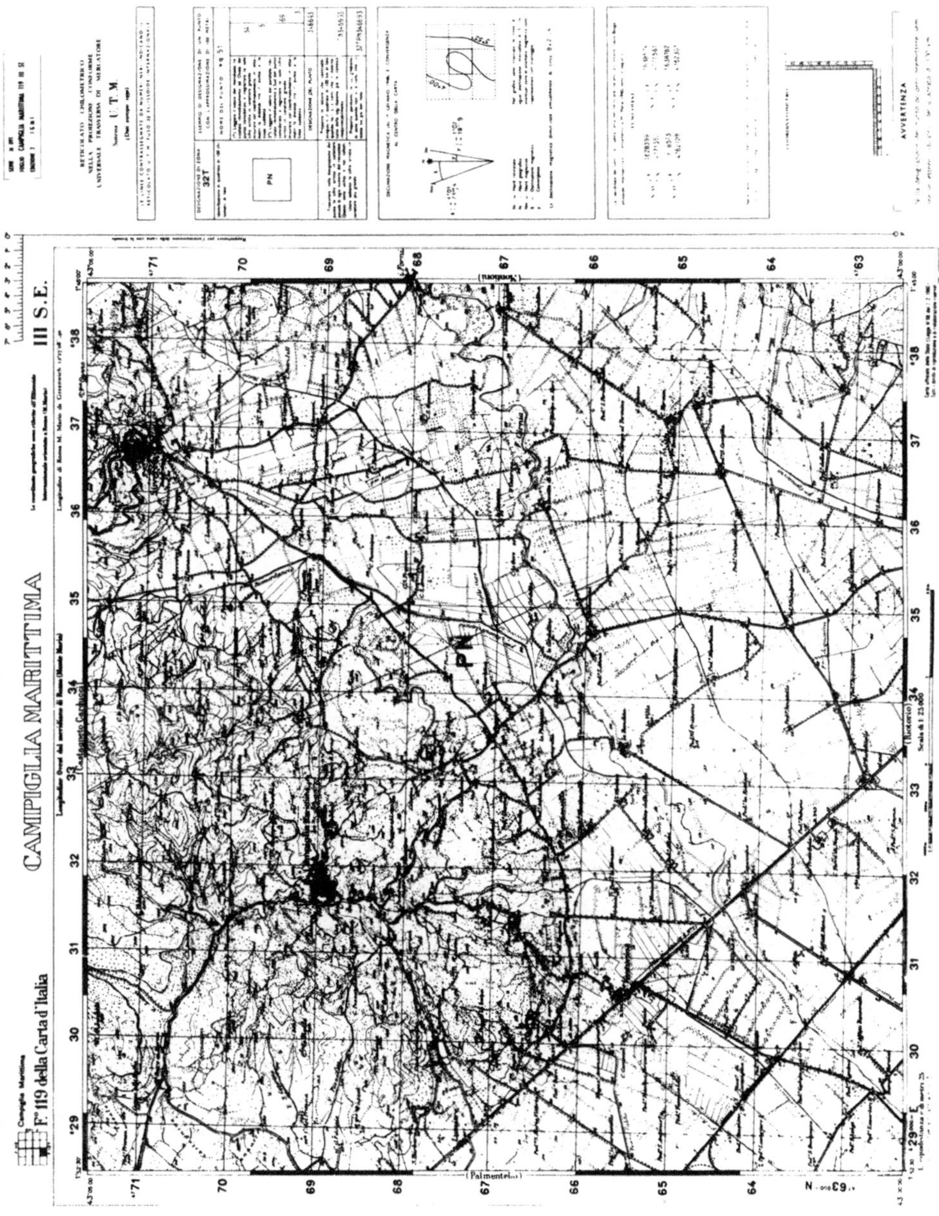
CAMPIGLIA MARITTIMA
F° 119 della Carta d'Italia
III S.E.
AVVERTENZA

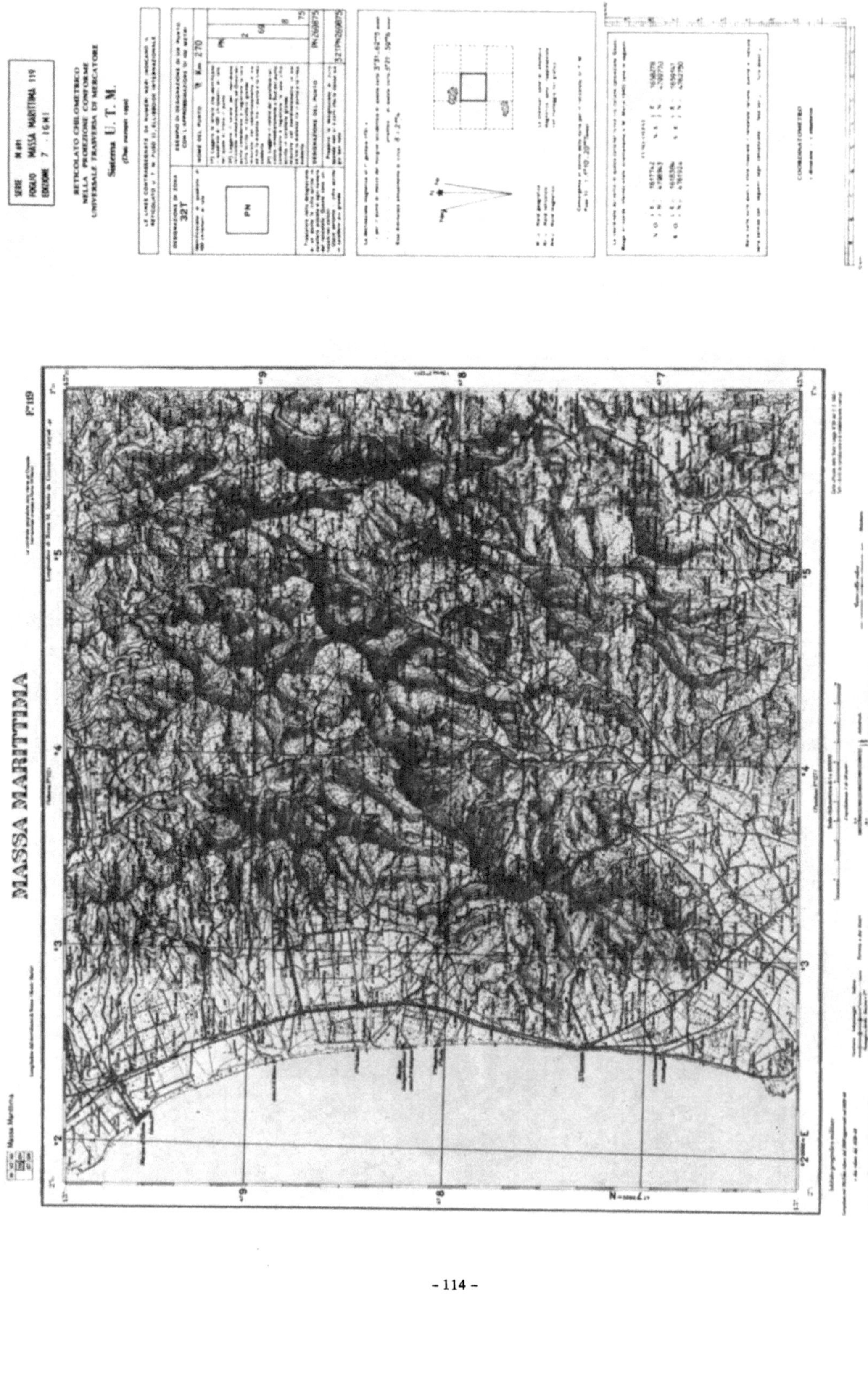

MASSA MARITTIMA
F° 119
SERIE M 691
FOGLIO MASSA MARITTIMA 119
EDIZIONE 7 - IGMI
RETICOLATO CHILOMETRICO
NELLA PROIEZIONE CONFORME
UNIVERSALE TRASVERSA DI MERCATORE
Sistema U. T. M.
32T
PN
COORDINATOMETRO

GASIFICATION OF WOOD IN THE CIRCULATING FLUIDIZED BED

METHANOL PRODUCTION ROUTE

Authors : Ch.Lindner, R.Reimert

Contract number : ESE/P/018/D

Duration : 18 months 1.Jan.1982 - 30.June 1983

Total budget : 1.681.976,- DM

CEC contribution : 840.988,- DM

Head of Project : R.Reimert, Lurgi Kohle und Mineralöltechnik GmbH

Contractor : Lurgi Kohle und Mineralöltechnik GmbH

Address : Lurgi Kohle und Mineralöltechnik GmbH
Bockenheimer Landstraße 42
D- 6000 Frankfurt/Main 1

Summary

By the gasification of wood in a circulating fluidized bed reactor a process is introduced exhibiting some advantages which derive from the special characteristics of the reaction system. Among other items the main features are high temperature with great constancy throughout the whole reactor allowing a tar free gas to be expected along with a long reactor lining lifetime and an easy scale-up which already had been demonstrated in an other application of the process principle.

In addition to the gasification the methanol synthesis can be demonstrated on-line in laboratory facilities (special service).

Taking into account the above and using the proposed test programme a set of data can be produced serving as a basis

- to establish performance data and
- economic evaluations,

for a commerical plant. Based on these results as well as on data about the biomass available investment decisions can be made with a high degree of confidence.

1. INTRODUCTION

Throughout the world there is a rising demand for energy making it appear increasingly more interesting to utilise alternative raw materials. Here a greater attention is paid to bio -masses, renewable raw material, and wastes from bio-masses. Wood is principally involved. In addition, regionally bark, straw and other wastes from bio-masses can also be of interest.

To evaluate the use of these wastes a plant concept is to be developed which produces a synthesis gas for methanol synthesis by gasification. As processes are offered on the market for gas purification and methanol synthesis, attention will principally be given to the development of the wood gasification process. The following targets are to be met by a commercial size plant:

- low investment costs
- simple operation
- high on-stream time
- gasification with a pressure of up to 30 bar
- throughputs up to 50 t/h and unit

To demonstrate wood gasification, a gasification reactor existing in the Lurgi laboratory and operating on the circulating fluidized bed (CFB) principle will be converted. Tests carried out with this pilot plant will provide all informations necessary for the design of a commercial size plant and for accompanying economic evaluations.

2. PROCESS PRINCIPLE OF THE "CIRCULATING FLUIDIZED BED"

2.1 General

The circulating fluidized bed can be placed in the transition range between the classical bubble-forming fluidized bed and the pneumatic transport reactor (Fig.1).

If gas flows through fine particle solid material with increasing velocity initially a fluidized bed of the classical form with defined surface is generated. It can be compared with a boiling liquid (condition A, Fig.1).

Here, only relatively few solid particles are entrained with the gas, separated in a cyclone and either discharged from the reaction system or recycled to the fluidized bed.

If the gas velocity increases, the fluidized bed expands until the solids are almost uniformly distributed over the reactor height. The quantity of solids entrained by the gas increases, is separated in the recycle cyclone and recycled to the reactor creating the circulating fluidized bed (B and C in Fig. 1). This leads to the recycle cyclone being a permanent constituent of the reactor system, which can be seen on Fig. 2 showing a calcinating reactor with two stage combustion.

Along with the external circulation inner recirculation of material takes place in the reactor due to constantly changing densely packed strands and clusters. This causes an extremely intensive mixing movement between gas and solids. In the diagram this fact is characterised by the rise of relative slip velocity between gas and solids. Together with the solids recycle inner circulation gives a uniform temperature across the overall circulating fluidized bed reactor.

The CFB operates advantageously in the range with maximum possible slip velocity at the given particle spectrum. This causes extremely good heat and mass transfer which enables a rapid heating up of feed material combined with an excellent reaction rate at constant temperature in the overall system of the circulating fluidized bed.

2.2 State of the art

The principles of CFB-reactors were introduced in alumina industry at first. There, in an endothermic reaction aluminium trihydroxide is dehydrated at a temperature of 1100 °C. Meanwhile 16 plants have been built or are under contract applying this "Lurgi/VAW-Fluid Bed Calcination" named process (s.Tab. I).

Using some laboratory and pilot scale plants Lurgi is inquiring into new fields of application of the CFB-principle whenever this sounds reasonable. As a result coal combustion became a second application of great importance.

Now investigations are directed to the gasification of wood in a CFB-reactor making use of its inherent advantages.

2.3 Features in case of wood gasification

The application of the CFB-principle for synthesis gas production via wood gasification provides the following advantages:

- No tar production due to
  - feeding the wood into the bottom part of the reactor resulting in a long residence time of the devolatilisation gas
  - rapidly heating up of the feed which partly suppresses the formation of higher hydrocarbons
  - the uniform temperature in the overall reaction system allows the gaseous phase to be kept on high temperature throughout its whole residence time.
- In view of the high gas velocity, a high specific throughput is achieved.
- Good part-load behaviour as it is possible to bring the

the load down into the range of the classical fluidized bed (abt. 20 % part-load).

- No danger of hot spots and sticking and clinkering.
- Gentle treatment and hence long life time of the brick material because of very constant temperatures.
- Easy scaling up.

3. PILOT PLANT FOR GASIFICATION

3.1 Description

For performing the gasification tests with wood one of the Lurgi owned pilot plants (s. Tab. I) will be used. As steam and oxygen will be used as the gasification agent some minor rearrangements are necessary. In a simple scheme Fig. 3 shows what the plant will then be like.

Wood, ground to an appropriate particle size and air dried, is fed by a cooled screw feeder into the circulating bed. The material is some inerts, sand or aluminium oxide for instance. Steam if necessary and oxygen are blown into the bed for gasifying the wood and its devolatilisation products. After de-dusting the raw gas in two cyclones and in a scrubber the gas can be used (e.g. for heating purpose).

The reactions take place under ambient pressure and at temperatures up to 1100 ^{o}C, these being the design data of the reactor. The use of existing equipment offers the possibility to obtain the data aimed for in a fast and inexpensive way. Techniques are available for the evaluation of the test results and their extrapolation up to a commercial sized reactor, i.e. greater dimensions and higher pressure.

3.2 Test objectives and programme

In general, the purpose of the tests is to establish a material and a heat balance for the process (s.Tab. II). Besides, investigations into the performance for different materials and temperatures as well as into the operational behaviour are on the programme.

With the help of some guidance, small scale tests, the up-scaling calculations will be done. On the basis of this the economic evaluation for a commercial size plant can be made finally.

On the attached block diagram (Fig. 4) the main activities of the programme and the estimated time schedule are shown. The advantage of making use of an existing plant is evident as no construction and installation time is necessary.

The budget of the programme will be shared by the CEC and by Lurgi in equal proportions.

3.3 Expected test results

On Fig. 5 the estimated balances for the pilot plant are shown. These figures, of course, must be proven.

Electric energy can be added to the heat balance. This form of energy, however, is merely used for drives and amounts in the pilot plant to 10 kWh/h.

Oxygen is delivered in liquid phase to the plant and will be evaporated. For oxygen production via air separation an energy consumption of appr. 0,8 - 1,0 kWh/Nm^3 can be assumed for a medium to big sized plant.

On Tab.III the expected performance data are given.

The data indicate a yield of 0,0475 kmol CO H_2 per kg of dried wood (20 % moisture). This will result in a methanol production of 0,40 kg per kg of wood as above.

4. METHANOL SYNTHESIS

As the subject of the research programme of the CEC is "Methanol from Wood" facilities for methanol synthesis are provided for also. All the necessary process-steps

- raw gas shift conversion
- gas purification
- methanol synthesis loop

are available in a laboratory scale. In Fig. 6, which shows the process scheme, it can be seen that a side stream of the product gas will directly be fed to the downstream units. This offers the possibility to demonstrate the whole chain of processes in one place.

However, it must be noticed that all the process steps downstream of the gasification reactor (gas cooling, dedusting, purification, shift conversion, synthesis, distillation) had been built by Lurgi on a commercial scale already (s. Fig. 7). The gases with which these units were fed, were similar to those expected from wood gasification or sometimes more difficult to handle due to their composition.

The work on methanolsynthesis will be done due to a special arrangement between CEC and Lurgi and is not covered by this project.

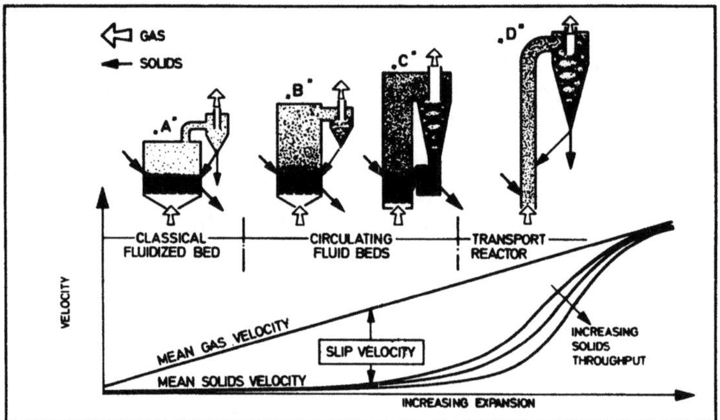

LURGI | Basic fluid bed combustion systems | FIG.1

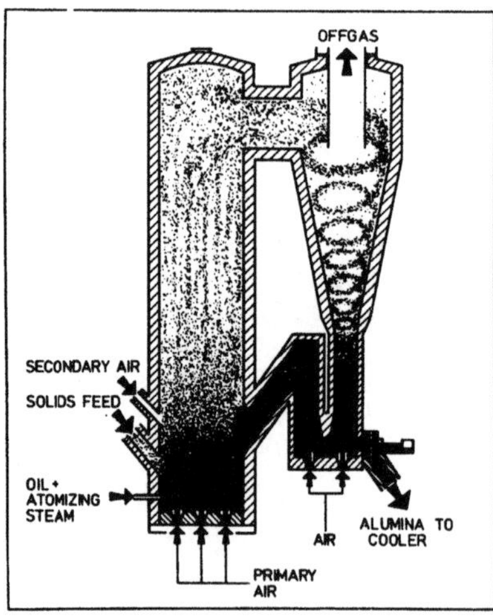

LURGI

TWO-STAGE COMBUSTION IN A CIRCULATING FLUID BED

FIG. 2

LURGI HAS BUILT OR UNDER CONTRACT

CIRCULATING FLUIDIZED BED

- 16 CALCINERS SINCE 1970

ϕ_{MAX} = 5 M, 1 BAR

– PILOT PLANTS
FOR GENERAL APPLICATIONS (TESTS)

ϕ = 0,4 M, H = 4,5 M, 1 BAR

ϕ = 0,7 M, H = 20 M, 1 BAR

– COAL COMBUSTION

ϕ = 4,5 M, 20 T/HR LOW GRADE COAL (50 % ASH)

H = 25 M, 1 BAR, N = 83 MW_{THERM}

HYDROGASIFICATION OF COAL (FLUIDIZED BED)

– PILOT - PLANT, 1975

ϕ = 0,2 M, 110 BAR, 200 KG / HR, 950 °C

– PILOT - PLANT, 1981 / 82

ϕ = 1,0 M, 120 BAR, 10 T / HR, 950 °C

LURGI | STATUS OF CFB-TECHNOLOGY REFERENCES | Tab. I

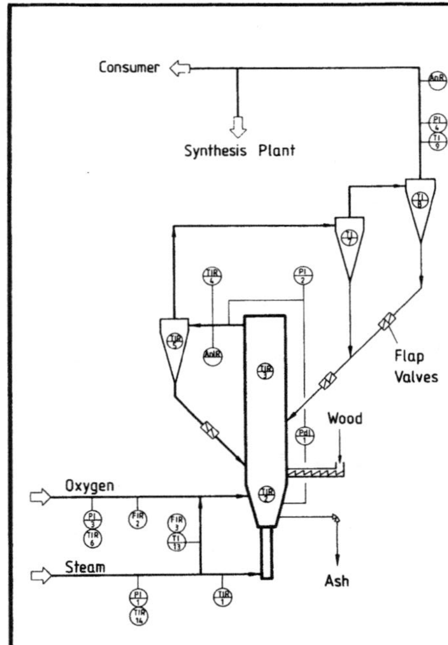

LURGI | Circulating Fluidized Bed Pilot Plant | FIG. 3

I HEAT AND MATERIAL BALANCE

- SPECIFIC CONSUMPTION FIGURES
  - STEAM
  - OXYGEN
- SPECIFIC PRODUCTION FIGURES
  - GAS, YIELD AND COMPOSITION
  - GAS LIQUOR

II OPERATIONAL BEHAVIOUR

- SPECIFIC REACTOR THROUGHPUT
- PARTIAL LOAD BEHAVIOUR
- COMPARISON OF CLASSICAL AND CIRCULATING FLUIDIZED BED

LURGI | OBJECTIVES FOR PILOT PLANT OPERATION | Tab. II

Gas composition

		dry	
CO	24,70	35,95	vol/vol
H_2	25,96	37,79	vol/vol
CO_2	17,11	24,9	vol/vol
CH_4	0,6	0,88	vol/vol
N_2	0,32	0,46	vol/vol
H_2S	0,01	0,02	vol/vol
H_2O	31,3	-	vol/vol
	100	100	vol/vol

Heating value (LHV): 8912 kJ/Nm^3 (dry)

Gas yield: 1,11 Nm^3 (dry) /kg (wood, 20 % moisture)

Cold gas efficiency (LHV): 78 %

Tab. III Expected Performance Data

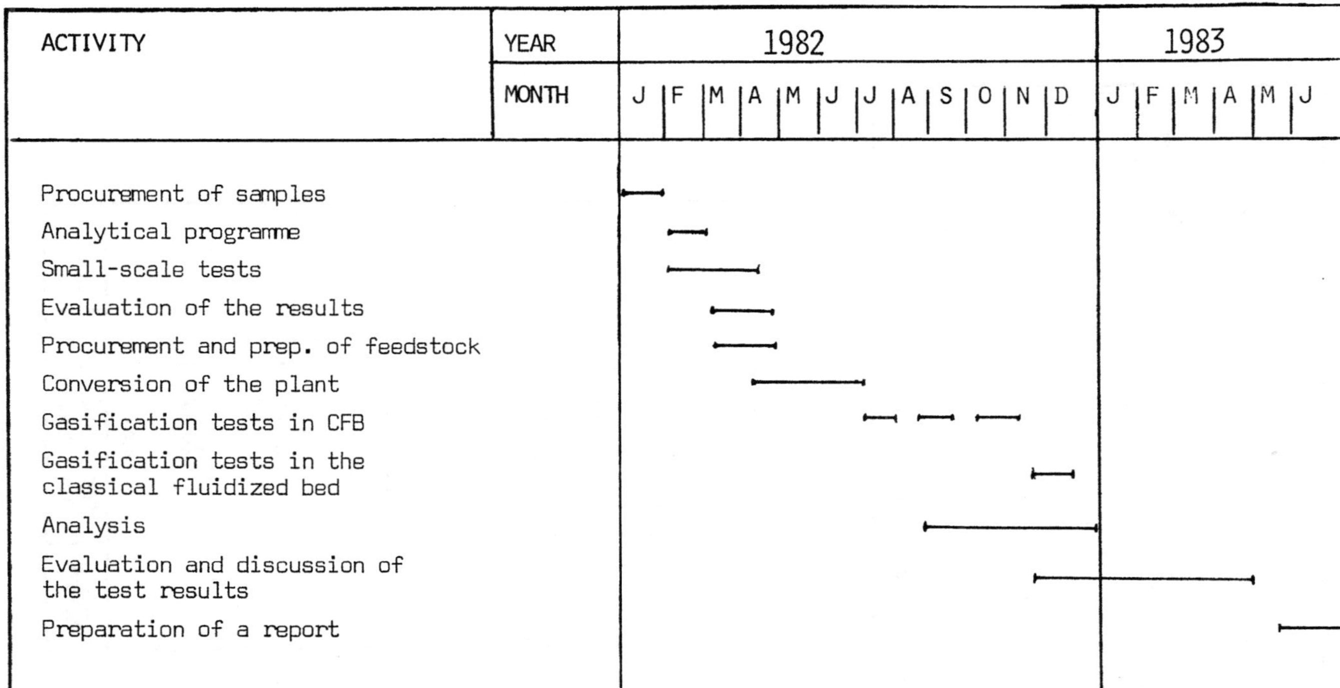

LURGI

TIME SCHEDULE

FIG. 4

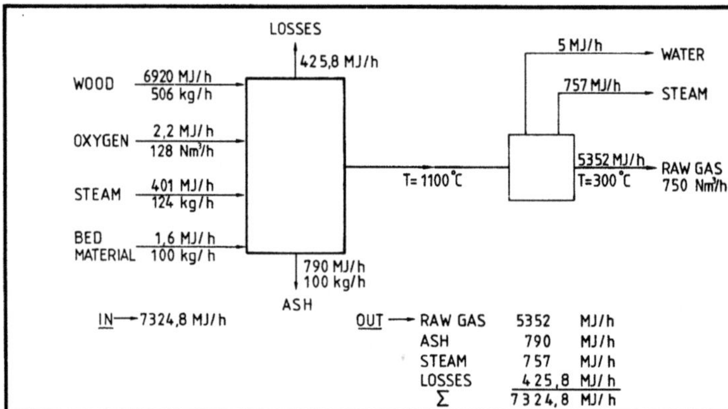

LURGI HEAT AND MATERIAL BALANCE OF PILOT PLANT FIG. 5

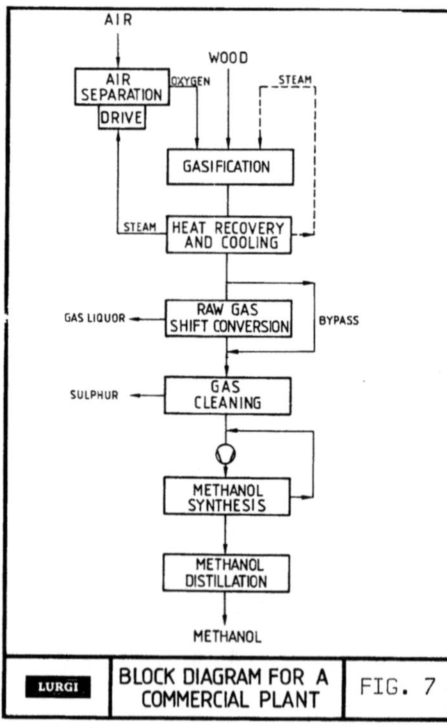

LURGI BLOCK DIAGRAM FOR A COMMERCIAL PLANT FIG. 7

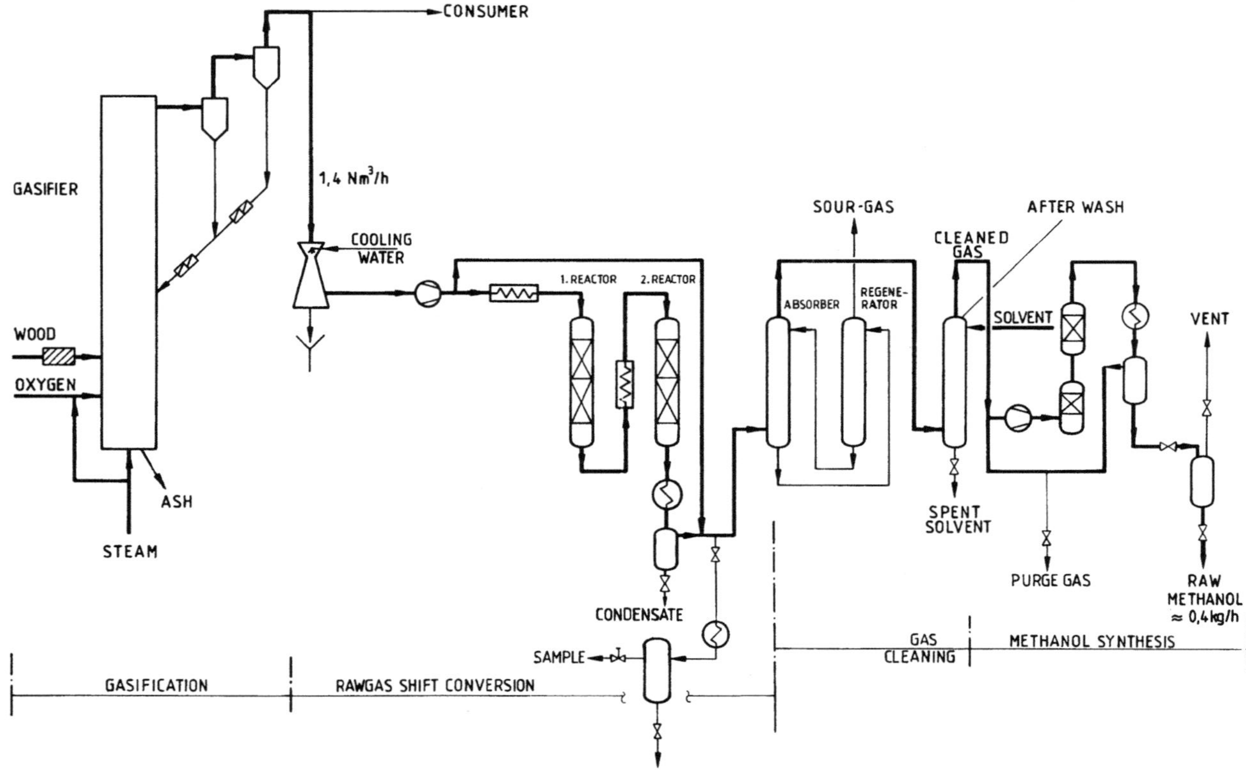

LURGI

LABORATORY METHANOL PLANT

FIG. 6

PART II - ALGAE

BIOTECHNOLOGIE SOLAIRE - PRODUCTION ET UTILISATION DES ALGUES

Author	: C. GUDIN
Contract number	: ESE/P/005/F
Duration	: 36 months 1 July 1980 - 30 June 1983
Total budget	: 8 730 000 F CEC contribution : 3 500 000 F
Head of project	: Département de Biologie, Service de Radioagronomie, Commissariat à l'Energie Atomique
Address	: Département de Biologie, Service de Radioagronomie Centre d'Etudes Nucléaires de Cadarache B.P. n°1 13115 Saint-Paul-lez-Durance, France

Summary

With a tubular culture system of 1 m^2 of photosynthesis surface, fully controled and automatized, we have been studying since 1976, in the South of France, the productivity of microalgae. The production in total dry matter reach 21 g/m^2/day (76 T/ha/year) on an annual average with a continuous production during 365 days. The energy balance, fixed energy over technology energy leads to no gain and must be improved. The cost of the biomass is between 4 and 8 FF/kg. A valuable target, in the actual state of the technology can only be a biomass with a market price > 10 FF/kg. This target exists with the microalga *Porphyridium cruentum* which in certain culture conditions we are controlingproduces 50% of its total biomass weight as a sulfated exocellular polysaccharide, soluble in the culture medium and precipitable with pure ethanol. This polysaccharide can be used as a thickening agent in food industries and presents interesting caracteristics for enhanced oil recovery.

After extraction of the polysaccharide (about 10 g/m^2/day or 36 t/ha/year), it remains an equivalent quantity of methanisable cellular biomass.

We propose a tubular culture system of 10 m^2, based on our technology which will be working during 1 year in 1983 and the conception of a pilot of 100 m^2 as a plan document in 1983. During the contract period we will continue to produce informations at the 1 m^2 level in order to improve the study of the market for the polysaccharide (food/drug/oil enhanced recovery alternatives).

1.1. INTRODUCTION

Dans le cadre du 1er contrat (1978), un capteur solaire biologique à double couche permettant la culture simultanée et séparée d'une microalgue et d'une bactérie photosynthétique a été réalisé à petite échelle. La culture algale utilise le bleu et le rouge orange du spectre solaire, la culture bactérienne sous jacente le vert, le rouge et l'infrarouge. La culture continue de ces deux organismes superposés permet de dépasser la limitation théorique actuelle de 126 tonnes/ha.an de biomasse sèche.

Toutefois, la source de protons de la culture bactérienne la moins onéreuse étant le méthanol pour les Athiorhodobactéries, il serait nécessaire pour des raisons économiques de s'adresser à des Thiorhodobactéries utilisant l'Hydrogène sulfureux, ce qui entraîne des problèmes technologiques de cultivateur non encore résolus.

Pour cette raison, le contrat à partir de 1979 a porté sur la culture monospécifique à une seule couche de microalgues dans des systèmes de culture tubulaire, au niveau de 1 m^2 de réception d'énergie lumineuse, pour un volume de culture de 30 litres. Ce système entièrement automatisé de culture continue a permis d'établir avec *Scenedesmus acutus* que la production moyenne annuelle pour la région méditerranéenne est, dans l'état actuel de la technologie, de 21 g/m^2/jour de Biomasse sèche (76 t/ha.an).

Une étude économique a montré que le coût de production, à une grande échelle, oscillerait entre 4 et 8 F/Kg de Biomasse sèche. Par ailleurs, le bilan énergétique dans l'état actuel de la technologie ne conduit qu'à un rapport énergie fixée/énergie utilisée voisin de 1. Dès lors, deux orientations s'imposaient :

1 - tout en s'appliquant à réduire les coûts de fabrication par un choix judicieux des matériaux et un bon contrôle des paramètres coûteux, viser des Biomasses à haute valeur commerciale (> 10 F le Kg).

2 - Améliorer la technologie, en faisant appel à des énergies renouvelables pour certaines opérations ou mieux encore, trouver des technologies nouvelles à moinsgrande dépense énergétique.

La première conclusion nous a amené à retenir des microalgues capables d'excréter des polysaccharides exocellulaires à haute valeur commerciale, susceptibles de devenir des produits de substitution des alginates, carraghénanes (produits de macroalgues ; 10 à 50 F/Kg). Le *Porphyridium cruentum* a été retenu.

La seconde conclusion nous a amené à nous orienter, au niveau de la recherche de laboratoire vers l'utilisation de cellules immobilisées dans un substrat plus ou moins transparent à la lumière et perméable aux échanges gazeux et liquides (Système continu à lit fixe). Un photoréacteur de 2 litres fonctionne ainsi depuis maintenant 20 mois avec des cellules de *Porphyridium cruentum* immobilisées dans une mousse de Polyuréthane et excrétant, sous certaines conditions, le polysaccharide directement dans le milieu de culture.

Avec cette nouvelle technologie le passage au développement est plus lointain car de nombreux problèmes restent à régler.

Les contraintes énergétiques et économiques de la culture contrôlée de microalgues nous ont amené à déboucher sur le concept relativement nouveau de la production de biomasse exocellulaire. Cette biomasse exocellulaire peut avoir une forte valeur marchande potentielle, c'est le cas des polysaccharides sulfatés de *Porphyridium* ou une haute valeur énergétique, c'est le cas des hydrocarbures de *Botryococcus braunii*. Sa production en culture continue impliquera une dépense relative d'énergie mais sa séparation ultérieure de la biomasse cellulaire, méthanisable, permettra de faire face à la fois à l'objectif économique (rentabilité de la culture) et à l'objectif production d'énergie (méthane)... tout du moins dans le cas de *Porphyridium cruentum*.

Si on transpose les productivités annuelles de *Scenedesmus acutus* à *Porphyridium cruentum* (les résultats obtenus le permettent) on peut atteindre une production annuelle de 30 t/ha de Biomasse cellulaire méthanisable à basse valeur commerciale et de 30 t/ha de Polysaccharides sulfatés ayant un prix de marché > 25 F/Kg.

Les analyses chimiques et essais divers, récemment effectués par la firme CECA sur notre produit, confirment les chances économiques d'une telle culture.

C'est sur cette base, et avec le concours de la CECA que nous proposons une étude d'un système de production pilote de 100 m^2 (le ARE/AN SOLAIRE) soigneusement préparée par une étude expérimentale de 1 an sur 1 m^2 puis de 1 an sur 10 m^2.

Le système de culture retenu est le système tubulaire déjà exploité au niveau de 1 m^2 en recherche (5 ans d'expérience et de résultats fiables et répétables).

La souche retenue est *Porphyridium cruentum* dont on maîtrise la culture dans notre système et qui fait par ailleurs l'objet de recherches en lit fixe sur mousses de polyuréthanes.

La méthode de récolte retenue est la centrifugation permettant la séparation de la biomasse cellulaire humide (ultérieurement méthanisable) de la biomasse exocellulaire hydrosoluble (le polysaccharide). Le polysaccharide est extrait de la phase aqueuse par précipitation à volume égal d'éthanol puis séché.

1.2. DESCRIPTION DU CULTIVATEUR EXPERIMENTAL ET DE L'EQUIPEMENT DE MESURES

Le cultivateur tubulaire de 1 m^2 a été décrit dans nos publications depuis 1978. Il se compose d'un récepteur solaire de 1 m^2 de photosynthèse ayant une occupation au sol de 6 m^2 et d'un mélangeur gaz/liquide constitué pour l'instant d'une colonne type "air lift". Une pompe centrifuge "excentrique" à débit électroniquement controlable permet la circulation continue entre le récepteur et le mélangeur. Sur ce circuit sont disposés des "capteurs physicochimiques" pour mesurer le pH, la température, le potentiel redox, l'O^2 dissous, le CO^2 dissous et la densité optique de la culture. A l'entrée et à la sortie de "l'air lift" la concentration de CO^2 dans l'air est constamment mesurée par un analyseur infrarouge. A la surface du récepteur solaire l'énergie solaire est mesurée en permanence à l'aide d'un pyranomètre. L'ensemble de ces mesures continues est accessible sur des enregistreurs pour un contrôle visuel immédiat du fonctionnement de la culture. Toutes les données sont centralisées sur un microprocesseur qui calcule en permanence des rendements de fixation du carbone et des rendements de Bioconversion de l'énergie solaire. Un contrôle et une régulation constante de la température de culture et de son pH permettent une grande stabilité et fiabilité des données. L'ensemble du système grâce à un circuit latéral alimenté par des micropompes doseuses permet d'obtenir la culture continue automatique. Par unité de temps une quantité x de milieu minéral est apportée en même temps qu'une quantité équivalente x' de culture est soutirée dans une colonne de récolte branchée sur une centrifugeuse qui permet la séparation des phases liquides et solides (Biomasse fraîche). Cette quantité x est fixée par le taux de dilution de la culture qui en principe est le plus proche possible du taux de croissance maximum de la culture de façon à ce que le temps moyen de résidence dans le cultivateur soit proche du temps de division cellulaire de la culture.

La productivité dépend donc de deux paramètres ; celui de l'aménagement optimal de l'espace (surface de faible épaisseur) et de celui de l'aménagement optimal du temps. (Par exemple une microalgue ayant la possibilité de se diviser toutes les 8 heures pourra sur un cycle de jour long de 16 heures produire de la biomasse deux fois dans la journée). La culture continue par l'ajustement du taux de dilution permet cette mise en phase du cycle lumineux et du cycle génétique cellulaire.

1.3. PROPOSITIONS DE RECHERCHES CONCERNANT UNE UNITE PREPILOTE DE 10 M^2 ET PILOTE DE 100 M^2

Il est bon de rappeler que la particularité essentielle de ce type de Biotechnologie Solaire est d'être bidimensionnelle, contrairement à la biotechnologie traditionnelle qui repose sur l'hétérotrophie, (la source d'Energie et de Carbone étant un composé carbonique hydrocarboné) et est tridimensionnelle.

Il découle de ce fait que le transfert de connaissances est plutôt à chercher du côté des "capteurs plans" thermosolaires que de la Biotechnologie traditionnelle (fermentations et bioconversions).

Par ailleurs, le caractère gazeux de la source de carbone (CO^2) implique le développement d'une technologie contrôlée des équilibres carbonés en milieu aqueux ($CO^2 \rightleftarrows HCO_3^- \rightleftarrows CO_3^=$) dans des systèmes continus à 3 phases (gaz/liquide/solide) à viscosité et troubles optiques variables (charge mi-cellulaire variable).

Le développement de cette technologie nécessite l'emploi des radioisotopes.

Les deux secteurs technologiques de support sont le "génie solaire" d'une part, le "génie biochimique" d'autre part.

Le système à développer est donc un système de surface (faible épaisseur de 3 à 10 cm) transparent à la lumière solaire, capable d'optimiser aussi bien l'effet thermique (maintien d'une température optimale pour la photosynthèse toute l'année) que photosynthétique (fixation du CO^2) de la lumière.

Il doit pouvoir être mis en place aussi bien sur des surfaces d'eaux calmes (flottant) que sur des sols non utilisables pour l'agriculture (tranchées, structures posées sur le sol, etc....).

La source de carbone est constituée par du CO^2 pur d'origine naturelle (gisements géochimiques) ou industrielle (unités pétrochimiques de type hydrocrackeurs ou cimenteries).

Le rapport surface de photosynthèse à la surface d'occupation au sol du système doit être amélioré. Au niveau du cultivateur expérimental de 1 m^2, il est de 1 pour 6 et doit tendre grâce à l'effet d'échelle pour 10 et 100 m^2 vers 1 pour 2, voire moins si possible.

1.3.1. Génie solaire

L'étude à faire doit porter en premier lieu sur les matériaux transparents pour la construction du système tubulaire en tenant compte des éléments suivants : prix au m^2, poids, durée de vie, contenu énergétique et spectre d'absorption. C'est ainsi qu'à l'heure actuelle, dans le cas du polyéthylène et avec une durée de vie fixée à 3 ans, le matériau représente 40 % du coût énergétique, 30 % des frais variables et 6 % du prix de revient au Kg de biomasse sèche produite. D'autres éléments d'appréciation au plan énergétique et économique sont disponibles dans les publications citées.

Une cible basée sur une prospective économique est donnée à titre indicatif. En supposant une EPR (efficacité Photosynthétique) de 70 TEP/ha an (Tonne Equivalent Pétrole par hectare et par an) un millier d'exploitations de 200 ha (soit 0,4 % du territoire terrestre français) permettrait de produire 14 millions de TEP d'ici 20 ans, soit la consommation française d'Hydrocarbures pour les transports routiers et aériens, dans la mesure où l'on saura fabriquer en très grande série (quelques centaines de millions de m^2 par an) des cultivateurs solaires dont le prix de revient serait inférieur à 100 F/m^2/an. Cette cible évidemment idéale et fictive donne une idée de la valeur qu'il faudrait atteindre au m^2 pour entreprendre par exemple, à grande échelle la production d'hydrocarbures à partir de *Botryococcus braunii*. Cette prospection technico-économique a été établie d'après certaines lois "d'économie de série" avec comme cibles : coût du m^2 inférieur à 100 F durée de vie de 10 ans pour un prix de vente des hydrocarbures de 5 F/Kg. Nous ne pouvons la détailler ici.

1.3.2. Génie Biochimique

Sur la base des résultats obtenus au niveau du m^2 on peut schématiser l'état actuel de la technologie comme suit :

Energie solaire + CO^2 fixé = Biomasse + O^2 produit
2610 Kcal/m^2/jour + 36 g/m^2/jour = 21 g/m^2/jour + 26 g/m^2/jour.

Dans cette équation simpliste mais réelle, le rendement calorique de la Biomasse est voisin de 4 % sur l'énergie solaire et pas très loin du Rendement maximum théorique situé à ≃ 6,6 %. Par contre, le CO^2 fixé ne représente que 10 % du CO^2 injecté au système (360 g/m^2/jour) et traduit bien l'effort à accomplir en ce domaine pour s'approcher de l'idéal, représenté par l'équation établie par Soëder.

$$6,14\ CO^2 + 3,65\ H_2O + NH_3 = C_{6,14}H_{10,3}O_{2,2}N + 6,85\ O_2$$

Dans les conditions actuelles de la technologie pour fixer 126 t/ha/an de CO^2 dans 76 t/ha/an de Biomasse sèche il faut dépenser 1260 t/ha/an de CO^2 pur, ce qui n'est pas une donnée économique négligeable.

On peut résumer le domaine d'étude de la façon suivante :

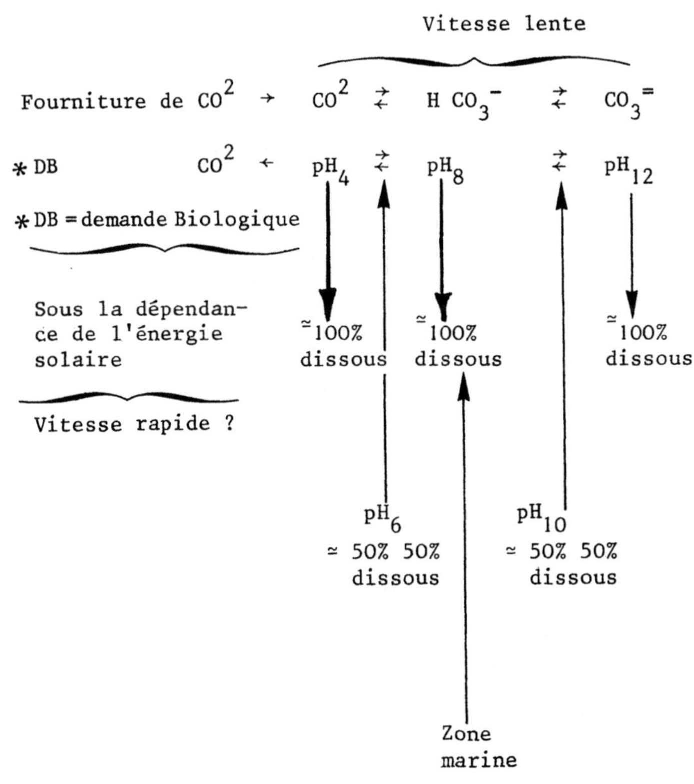

Selon le pH de la suspension de la microalgue, les réactions d'échanges CO_2, bicarbonate, carbonate peuvent constituer un facteur limitant pour la capture photosynthétique du CO^2. Il en résulte la possiblité d'une photorespiration importante dont on sait qu'elle dépend de la compétition "in situ" du CO^2 et de l'O^2. D'où l'intérêt de préciser d'une part les cinétiques d'échange isotopique CO_2 - H_2O en fonction du pH, d'autre part les valeurs brutes des processus = photosynthétiques et photorespiratoires en faisant dans les 2 cas appel aux indicateurs nucléaires (en l'occurence 13_C et 18_O) qui représentent la seule méthode de détermination des flux unidirectionnels.

L'objectif est de définir les niveaux optimum de CO_2 et le cas échéant d'O_2 dissous permettant d'aboutir à une photosynthèse maximale et une photorespiration minimale.

.4. DESCRIPTION DU PROJET PREPILOTE (10 M^2) ET PILOTE (100 M^2)

La description ci-après se limite au cas de la production de polysaccharides par *Porphuridium cruentum*.

Quelle que soit l'échelle prise en compte (1, 10 ou 100 m^2) le système comporte 3 parties :

1. Culture de l'algue, avec automatisation des contrôles, analyses et régulations.

2. Acquisition et traitement des données, conduisant à une modélisation.

3. Récolte et séparation de la biomasse produite en une phase solide destinée à la méthanisation (biomasse cellulaire) et une phase liquide plus ou moins visqueuse à traiter par l'éthanol pour extraction des polysaccharides (biomasse exocellulaire).

C'est essentiellement pour la partie culture que les problèmes d'échelle se feront sentir tant au niveau de la conception qu'à celui de l'exploitation.

Le passage successif aux échelles supérieures (de 1 à 10, puis 100 m^2) va requérir en effet des études de génie solaire, chimique et biologique. Signalons en particulier :

- l'optimisation du matériel support, en fonction du prix, de la durée, des qualités optiques, mécaniques et thermiques,
- l'optimisation des facteurs nutritifs,
- le transfert matière en milieu liquide, notamment l'alimentation carbonée (cinétiques d'équilibre CO_2, HCO_3^-, CO_3^{--}) par rapport à la demande biologique,
- la mise au point des processus de fractionnement de la biomasse,
- l'analyse et l'homologation du produit.

Les opérations se dérouleront comme suit :

A. Exploitation du cultivateur de 1 m^2 (échelle de laboratoire) : 1981-1982

Le système déjà décrit et mis au point dans le cadre du contrat de recherche 1978-1981 sera utilisé après adaptation à la culture de *Porphyridium cruentum*.

Son exploitation sera réalisée sur une période de longue durée (12 mois) dans le but d'évaluer et d'optimiser les performances de cette microalgue, et de procéder aux expérimentations et essais requis pour la conception de l'étape ultérieure (10 m^2).

N.B. : Si l'état d'avancement des études préalables concernant le *Botryococcus* le permet (Cf. : collaboration avec le CNRS) une expérimentation pourra être proposée à l'échelle du m^2 pendant l'exercice 1982-1983.

B. Conception, réalisation et exploitation du cultivateur de 10 m^2 (échelle pré-pilote) : 1981-1983

Il existe 3 options possibles pour le cultivateur :

1) système tubulaire posé au sol dans une serre thermostatée ;
2) système tubulaire terrestre à l'extérieur avec deux sous-options :
   a) tranchées imperméabilisées couvertes d'une enveloppe plastique
   b) tubulures transparentes à même le sol.
3) système tubulaire flottant sur une surface d'eau.

après expérimentations portant sur plusieurs variantes réalisées à des échelles intermédiaires.

A l'intérieur de ces 3 options, la comparaison -pour ce qui concerne le génie chimique (transferts du gaz et circulation de liquides)- entre systèmes mono et bi-phasiquessera expérimentalement établie, de même que l'on étudiera différentes solutions pour l'alimentation en CO_2 : injection par "mélangeurs de gaz" en 1 ou plusieurs points du circuit tubulaire, alimentation à contre courant, toutes ces études ayant pour objet de définir les solutions à retenir pour l'échelle de 10 m^2.

Après réalisation, le cultivateur de 10 m^2 sera expérimenté pendant 12 mois avec *Porphyridium* pour essais, optimisation et production de quantités suffisantes de biomasse cellulaire et exocellulaire.

L'optimisation de cette production requerra l'acquisition d'un viscosimètre à mesure continue, la production du polysaccharide se traduisant par une augmentation de la viscosité du milieu.

Le fractionnement en biomasse cellulaire et exocellulaire s'effectuera par centrifugation en continu.

Une certaine quantité de biomasse cellulaire fraîche sera stockée au froid, puis expédiée chaque mois à LOUVAIN pour études de méthanisation.

La quantité de biomasse exocellulaire produite sera suffisante pour des études poussées permettant la caractérisation et l'homologation du produit par l'intermédiaire de la CECA.

C. Conception du cultivateur de 100 m^2 (échelle pilote) : 1982-1983

On se propose d'étudier, au vue des résultats acquis à l'échelle de 1 m^2 (81-82) et 10 m^2 (82-83) l'option technologique la plus favorable, au double point de vue énergétique et économique, pour une unité pilote de 100 m^2. Le projet sera mené jusqu'au stade d'un dossier-plan aussi détaillé que possible.

2. CONCLUSIONS

Les recherches entreprises depuis 1976 et faisant l'objet d'un contrat dans le cadre de la CCE depuis 1978 ont abouti à la construction d'un cultivateur tubulaire de 1 m^2 de surface de photosynthèse complètement automatisé et entièrement contrôlé basé sur l'aménagement de l'espace (surface) et du temps (mise en jeu de la culture continue) ce qui permet d'approcher une Bioconversion de l'Energie solaire théoriquement fixée à 6,6 %. Le rendement moyen annuel est de 4 %, la productivité moyenne de 21 g/m^2/jour (76 t/ha/an) mais le rendement d'utilisation du CO^2 fourni n'est que de 10 % et doit être amélioré. Lors ce ces recherches une étude technicoéconomique a montré que les coûts de production de la biomasse sèche se situaient entre 4 et 8 F/Kg et qu'il fallait viser des Biomasses vendables à > 10 F/Kg pour rentabiliser la culture.

Pour l'instant, le seul créneau économique correspondant à cette contrainte est un polysaccharide sulfate proche des carraghénanes dont la valeur marchande dépasse 25 F/Kg.

Une microalgue rouge *Porphyridium cruentum* est capable d'excréter un tel produit sous forme de Biomasse exocellulaire dans le milieu de culture. Sa récolte en est possible par précipitation à l'éthanol.

Nous nous proposons donc, tout en continuant pendant un an (1981-1982) l'optimisation de cette production au niveau du 1 m^2 de construire un prototype prépilote tubulaire de 10 m^2 qui fonctionnera en (1982-1983). L'ensemble des informations biologiques et technicoéconomiques nous permettra fin 1983 de proposer au niveau d'un dossier plan aussi détaillé que possible un prototype pilote de 100 m^2.

Les productions espérées au niveau du 100 m^2 (le ARE/AN SOLAIRE) sont de l'ordre de 300 kg de polysaccharide exocellulaire à 25 F le Kg et de 300 Kg de biomasse cellulaire (poids sec) transformables en CH_4 par an.

Une unité (ultérieure) industrielle de 1 hectare (100 m/100 m) conduirait à une production de 30 tonnes de polysaccharide ce qui pour ce genre de marché correspondrait à une unité de taille suffisante. La taille du marché mondial des carraghénanes est de 10.000 t/an, celle de la production Française de 2000 tonnes an et ce nouveau polysaccharide dans la mesure de l'autorisation de son homologation couvrirait un créneau commercial compatible avec de tels tonnages.

Par ailleurs, les 30 tonnes de Biomasse cellulaire sèche provenant de cette bioconversion constitueraient un sous-produit non négligeable pour la méthanisation.

Les études technicoéconomiques réalisées actuellement dans le monde, même sur des systèmes de culture en apparence plus simples et moins coûteux, ne permettent pas pour l'instant une rentabilité de culture basée sur la production d'énergie, qu'il s'agisse de méthane ou d'hydrocarbures.

La conception d'une unité industrielle rentable dans les 5 ans à venir passe nécessairement par la production d'un produit à haute valeur spécifique.

Ce raisonnement n'est valable que dans le cas d'un pilote avec une application à court terme (5 ans). D'autres voies de "rentabilisation" de la production d'énergie par bioconversion cellulaire existent mais nous sommes là dans le domaine de la recherche à moyen terme (10 à 15 ans). C'est le cas par exemple de la production directe de biomasse exocellulaire dans des Bioréacteurs à lit fixe par la technique d'Immobilisation et de stabilisation des cellules sur des supports transparents à la lumière. Cette technique pourra peut-être s'appliquer par exemple à la production d'Hydrocarbures avec *Botryococcus braunii*.

B I B L I O G R A P H I E

Patents

- Claude Gudin et Eric Peel - 1975
  Method for growing undifferentiated plant tissue. UK Patent 1, 401, 681.

- Claude Gudin - 1976
  Procédé pour faire croître des cellules de plantes. Brevet français 2, 304, 277.

- Claude Gudin - 1976
  Method of growing plant cells. U.S. Patent 3, 955, 317.

- Claude Gudin - 1978
  Procédé et dispositif pour le développement et la culture de matières photosynthétiques en suspension dans un milieu aqueux. Brevet français, 2, 361, 060.

- Eric Peel, et Claude Gudin - 1974
  Procédé pour produire la croissance de plantes et plantes obtenues. Brevet français 2, 186, 186.

European Commission Reports

- Claude Gudin - 1978
  Development of a solar captor with two photosynthétic layers for storage of energy fixed by microalgae and bacteria suitable as energetic, chemical or feeding sources. (17 pages) 2nd Coordination meeting - 9-10 nov. 1978 Brussels.

- Claude Gudin - 1979
  3rd Coordination meeting. 6-7 june 1979 Taormina (42 pages).

- Claude Gudin & Daniel Chaumont - 1980
  For a Biotechnology based on microalgae (33 pages) 4th Coordination meeting 17-19 sept. 1980, Amsterdam.

- Claude Gudin & Daniel Chaumont & Odile Desanti - Daniel Pioline. 1981.
  5th coordination meeting. 22-24 june 1981 (5 pages + reprints). Copenhague.

Publications

- Claude Gudin 1976
  Bioconversion of solar energy. p.48-51 in Solar Energy in Agriculture, sept. 1976 Reading Publication UKI ISES.

- in Heliosynthèse et aquaculture Martigues sept. 1978. édité par le CNRS (311 pages).

- Jean-Paul Braud (CECA)
  l'Industrie française des phycocolloïdes (CECA) p.43-56.

- Claude Gudin
Production de Biomasse à partir de microphytes p.59-62.

- Claude Gudin et Daniel Chaumont
Bioconversion de l'Energie solaire dans un système double couche (microalgues/bactéries photosynthétiques) p.71-84.

- Claude Gudin
Place des cellules végétales dans la production de biomasse, p.117-118.

- Daniel Thomas, Marie-Françoise Cocquempot et Véronique Larreta Garde, Immobilisation et stabilisation de thylakoïdes et de chromatophores, p.223-244.

- Claude Gudin, Daniel Chaumont, Odile Desanti, Daniel Pioline. 1981
Culture continue d'Organismes cellulaires chlorophylliens pour des productions spécifiques (p.147-166) dans Revue du Palais de la découverte n° 21 de juillet 1981 "Les Bases Scientifiques de l'Amélioration des Ressources Alimentaires" (99ème congrès de l'A.F.A.S. à Amiens 8-12 sept. 80.).

- Claude Gudin, Daniel Chaumont, Odile Desanti et Daniel Pioline. 1980
Pour une biotechnologie solaire basée sur les microalgues. p.55-62. Revue internationale d'Héliotechnique. COMPLES 2ème semestre 1980.

- Claude Gudin & Daniel Chaumont. 1980
A Biotechnology of photosynthetic cells based on the use solar Energy. p481-482, Biochemical Society Transactions 8 (4).

- Claude Gudin & Daniel Chaumont & Elisabeth Berra. 1980.
New concepts in Solar Biotechnology in Energy from Biomass. Brighton, nov. 1980.

- Elisabeth Berra. Juin 1980.
Immobilisation de cellules de Porphyridium cruentum, microalgue excretant un polysaccharide. Projet de Fin d'Etudes (56 pages, 21 annexes illustrées).
Université de Technologie de Compiègne.

- Brigitte Thomasset. 17 juin 1981.
Etudes Cinétiques et morphologiques de cellules et d'Organelles immobilisées.
Thèse de Docteur Ingénieur. (177 pages, 187 références, 17 planches photographiques)
Université de Technologie de Compiègne.

- Claude Gudin et Daniel Thomas. Note présentée par Raymond Latarjet. 1981
Production de polysaccharides sulfatés par un biophotoréacteur à cellules immobilisées de Porphyridium cruentum (3 pages).
Comptes-rendus de l'Académie des Sciences, Biochimie Appliquée III - 293- le 6 juillet 1981 (sous presse).

- Claude Gudin. 1981
Biomasse algale et biotechnologie solaire p.394-396 in "La Pêche Maritime" n° 1240 juillet 1981. (Conférence de l'ANEL, 30 oct. 1980, Fréjus).

- Claude Gudin. 1981
La Biomasse Algale. in Cahier de l'A.F.E.D.E.S. n° 6. Photosynthèse. Energie-Biomasse 224 p. 54 fig. 45 tabl. Editions Européennes Thermique et Industrie, 30, rue de la Source 75016 PARIS, à paraître en sept. 1981.

Selections of books or reviews relative to the heliosynthesis and to solar biotechnology

- Le pétrole vert. Caroline Cosse Manière, 1981. Editions Apogée (95 p)., p.16-21.

- Science et Vie n° 765. juin 1981, p.113.

- Solaire 1 Magazine, déc/janv. 1981 n°12, p.8-10.

- Environnement et Cadre de vie. Recherche, les nouvelles frontières, n° 1550, janv. 1981 p.33-34.

- L'Energie Verte. Livre du COMES 1980. Voir chapitre sur Biomasse Algale et Bioconversion directe de l'énergie solaire.

- Production de molécules à haute valeur ajoutée à l'aide de chromatophores immobilisés. Véronique Larreta Garde. 16 octobre 1981. Thèse de docteur Ingénieur U.T.C.

- Bioconversion directe de l'énergie solaire par Spirulina maxima en culture continue.
Daniel Chaumont 1981 - DEA - UTC.

- Contribution à l'étude Physiologique de Porphyridium cruentum immobilisé sur mousse de Polyuréthane - Catherine Thepenier 1981 - DEA - UTC.

- Biophotoréacteur à cellules immobilisées de Porphyridium cruentum - Claude Gudin - Daniel Thomas - Catherine Thepenier - Daniel Chaumont 1981- Brighton in "Solar Wordl Forum" organisé par l'I.S.E.S. (sous presse) 10 pages.

- A new approach to the study of aqueous solutions of CO_2 - R. Gerster. T. Maren - D - Silverman Proc of the 1st int. conf. on stable isotopes in chemistry, Biology and medicine, Argonne - 1973 pp 219 - 228.

- Oxygen as a tool for studying photorespiration O_2 uptake and incorporation into glycolate, glycine and serine. R. Gerster. - B. Dimon. - P. Tournier - A. Peybernes. - C.R. du congrès sur les applications des isotopes stables. Leipzig 1978.

- Analysis of photorespiratory oxygen uptake on various plant species using ^{18}O. - P. Jolivet, C. Triantaphylides, G. Peltier, R. Gerster. - P. Thibault forth intern. Congress of Photosynthesis, Sept. 1980.

HYDROCARBON PRODUCTION VIA CULTIVATION OF THE ALGA BOTRYOCOCCUS BRAUNII

Authors :	E. CASADEVALL, C. LARGEAU.	
Contract number :	ESE-P-011-F	
Duration :	24 months	1 July 1981 - 30 June 1983
Total budget :	F 899.000	CEC contribution : F 449.000 (50 %)
Head of project :	Dr. E. CASADEVALL, Directeur de recherche.	
Contractor :	Ecole Nationale Supérieure de Chimie de Paris.	
Address :	E.N.S.C.P. Laboratoire de Chimie Bioorganique et Organique Physique 11, rue P. et M. Curie 75231 PARIS CEDEX 05	

Summary :

The green unicellular alga Botryococcus braunii with a hydrocarbon content as high as 75 % of the biomass dry weight, holds a good place among plants with a large energetic potential. The observation of the contemporary species in nature, and studies on the fossil species largely implied in petroleum origin, underline the high productivity of this alga. The results of ours recent investigations about effects of culture parameter variations on hydrocarbon production which show that, in culture, hydrocarbon productivity can reachs 15 g/m^2/day now support the feasability of an industrial production of "solar hydrocarbons" via a large scale cultivation of this alga.

The project related to B.b. cultivation which is introduced, enters in the frame of a collaboration with the CEA team. It is aimed to the design and the experimentation of a laboratory micropilote. The purpose of this experimentation is to test B.b. cultivation in conditions near those used in the 1,10 and 10 m^2 scale cultivators which are designed by the CEA team for Porphyridium cruentum (P.c.). The experimentation in this micro-pilote will allow :

- to detect the specific problems of B.b. culture and to find solutions so as to adjust more easily to B.b. cultivation the 1, 10, 100 m^2 scale pilot units built for P.c.
- to evaluate the cost of recovered hydrocarbons.
- to obtain adequate hydrocarbon supply for investigating the possible channels of B.b. hydrocarbon valorization.

PRODUCTION D'HYDROCARBURES PAR CULTURE DE L'ALGUE BOTRYOCOCCUS BRAUNII

Auteurs : E. CASADEVALL, C. LARGEAU.

Contrat numéro : ESE-P-011-F

Durée : 24 mois 1° Juillet 1981 - 30 Juin 1983

Total budget : F. 899.000 CEC contribution : F 449.000 (50 %)

Chef du projet : Dr. E. CASADEVALL, Directeur de recherche

Contractant : Ecole Nationale Supérieure de Chimie de Paris

Adresse : E.N.S.C.P.
Laboratoire de Chimie Bioorganique et Organique Physique
11, rue P. et M. Curie
75231 PARIS CEDEX 05

RESUME :

L'algue verte unicellulaire Botryococcus braunii (B.b.) avec un contenu en hydrocarbures qui peut atteindre jusqu'à 75 % du poids de biomasse sèche occupe une bonne place parmi les plantes à haut potentiel énergétique. Les observations relatives à l'espèce contemporaine dans son milieu naturel, ainsi que l'étude de l'espèce fossile largement impliquée dans la genèse des pétroles, suggèrent que cette algue a une productivité élevée. La faisabilité d'une production industrielle d'hydrocarbures"solaires" par la culture à grande échelle de B.b. est de plus étayée par les résultats de nos récentes recherches qui montrent que dans des conditions de culture appropriées, il est possible d'atteindre une production d'hydrocarbures de 15 g/m^2/jour.

Le projet présenté ici entre dans le cadre d'une collaboration avec l'équipe du CEA. Il a pour objectif la conception et la réalisation d'un micro-pilote de laboratoire. Ce micro-pilote est destiné à expérimenter la culture de B.b. dans des conditions proches de celles adoptées par l'équipe du CEA pour les unités pilotes de 1, 10 et 100 m^2 qui seront d'abord utilisées pour la culture de Porphyridium cruentum (P.c.). L'expérimentation du micro-pilote permettra :

1) de détecter les problèmes propres à la culture de B.b. et de leur trouver des solutions de façon à faciliter l'adaptation à la culture de B.b. des unités pilotes de 1, 10 et 100 m^2 mises au point pour P.c.
2) de faire une première évaluation du coût des hydrocarbures produits.
3) d'obtenir des quantités suffisantes d'hydrocarbures pour effectuer une étude approfondie des différentes filières de valorisation de ces composés.

1.1 Introduction - Le projet que nous présentons a pour objectif l'expérimentation à l'échelle pilote d'une culture de Botryococcus braunii comme source renouvelable d'hydrocarbures. Ce projet comporte dans un premier temps la conception et la réalisation d'un appareil qui doit être considéré comme l'unité de base utilisable en laboratoire pour la culture de Botryococcus. Cet appareil sera conçu comme un élément du cultivateur de 1 m^2 décrit par GUDIN et coll. Comme ce cultivateur, il sera équipé de systèmes de mesure et de régulation des paramètres physico-chimiques de la culture et pourvu d'un système de traitement automatique des données. Il permettra de réaliser le premier stade de l'expérimentation. Il est destiné à acquérir un complément d'informations sur les besoins spécifiques de la culture de B.b. de façon à ce que les essais au niveau du cultivateur de 1 m^2 puissent être menés avec de bonnes chances de succès. D'utilisation plus souple que ce dernier, il en facilitera la mise au point.

L'expérimentation au niveau du cultivateur de 1 m^2 sera effectuée en collaboration avec l'équipe du C.E.A. (C. GUDIN) sur l'appareil conçu et réalisé par cette équipe pour la culture de Porphyridium cruentum. Lorsque les problèmes,qui eventuellement se poseraient au niveau de ce cultivateur auront été résolus, l'expérimentation se poursuivra successivement sur les cultivateurs pilotes de 10 m^2 et 100 m^2, dont l'étude et la mise au point font l'objet du co-projet de l'équipe du C.E.A.

Justification du projet - L'algue verte unicellulaire Botryococcus braunii (B.b.) se caractérise par un contenu en hydrocarbures (figure I) qui peut atteindre jusqu'à 75 % du poids sec de biomasse. Bien que la présence d'hydrocarbures ait été signalée dans beaucoup d'organismes photosynthétiques, cette espèce est, à notre connaissance, la seule pour laquelle des taux aussi élevés ont été rapportés.

Des études géochimiques en relation avec la formation des charbons d'algues et la genèse des pétroles montrent par ailleurs que l'espèce Botryococcus a donné lieu au cours des temps géologiques à des proliférations remarquables. Plus près de nous, des blooms spectaculaires de la forme actuelle confirment qu'il s'agit d'une espèce ayant encore une grande productivité. Sa culture pouvait donc être envisagée comme un moyen de remplacer directement, au moins en partie, les "hydrocarbures fossiles" en voie d'épuisement par des "hydrocarbures solaires" constamment renouvelés. Le fait qu'il s'agisse d'une micro-algue était une raison supplémentaire de penser que cette idée avait des chances de pouvoir un jour se réaliser. En effet il est bien connu que les algues unicellulaires sont parmi les organismes photosynthétiques ceux qui assurent les meilleurs rendements en biomasse. De plus, organismes aquatiques, leur culture n'entre pas en compétition avec les cultures alimentaires pour l'occupation des sols. Lorsqu'en 1976 nous avons entrepris l'étude de B.b. afin de tester la faisabilité d'un projet de production d'hydrocarbures par culture à grande échelle de cette algue, on notait dans la littérature, non seulement l'absence presque totale d'informations concernant sa physiologie, son métabolisme, son ultra-structure, mais aussi le manque de travaux se rapportant à sa culture.

Depuis cette date, les travaux que nous avons effectués nous ont permis de recueillir un nombre important d'informations qui viennent à l'appui de notre projet. Nous rapportons ici celles qui nous paraissent les plus propres à en faire ressortir l'intérêt.

Ainsi, nous avons montré que les hydrocarbures de B.b. se forment et se stockent principalement dans la paroi externe de l'algue. Grâce à cette localisation externe on peut donc envisager une récupération facile des hydrocarbures produits (figures II et III).

Concernant la culture de B.b., les premières expériences que nous avons effectuées dans des conditions de culture "standard" en batch (milieu minéral dilué, sans aération ni agitation) ont confirmé un fait signalé dans la littérature, à savoir que le développement en culture de cette algue est lent (temps de doublement de la biomasse de 1 semaine environ) en dépit d'un rendement photosynthétique normal. Dans ces conditions (conditions I du tableau I) le taux d'hydrocarbures rapporté à la biomasse est proche de la valeur minimale observée dans la nature. Toutefois, avec le même milieu minéral, mais en modifiant les conditions de culture, nous avons observé que par un ajustement convenable de ces conditions, on peut accroître de façon importante les rendements en biomasse et en hydrocarbures (conditions 6 et 7, tableau I). On obtient alors une courbe de croissance (figure IV) qui fait apparaître des temps de doublement de la biomasse de 2 jours 1/2 pendant la phase exponentielle.

TABLEAU I - Influence des conditions de culture sur la productivité

Conditions de culture	Biomasse sèche g/l	Hydrocarbures g/l	Hydrocarbures % du poids sec	Hydrocarbures productivité g/l/j
1	2	0.34	17	0.011
2	6	1.77	29.5	0.084
3	1.33	0.24	18	0.080
4	1.33	0.21	16	0.053
5	0.71	0.10	14	0.050
6	5.3	1.50	29	0.088
7	7	2.52	36	0.148

Conditions de culture : 1 culture en batch, non aérée, non agitée, température 20°C, récolte après 30 jours. 2 culture en batch, aérée (air + 1 % CO_2) agitée, température 26°C, récolte après 21 jours. 3 mêmes conditions que 2, récolte après 3 jours. 4 culture continue mêmes conditions que 2, temps de renouvellement 4 jours. 5 identique à 4, mais temps de renouvellement de 2 jours. 6, culture en batch avec les conditions de 2, récolte après 17 jours. 7 identique à 6, mais concentration initiale en nitrate X2.

Ces résultats, qui soulignent qu'à de faibles variations des conditions de culture correspondent d'importantes variations de la production d'hydrocarbures, montrent que cette production est donc susceptible d'être optimisée. Les données des expériences précédentes (comparaison de la 2ème et de la 3ème ligne du tableau I) mettent par ailleurs en évidence que la production d'hydrocarbures peut être élevée, non seulement pendant la phase de repos (ce qui est généralement observé pour d'autres algues riches en lipides), mais aussi pendant la phase de croissance exponentielle. Pour confirmer cette information, une étude systématique a été menée de façon à déterminer l'effet de l'état physiologique de B.b. sur la production d'hydrocarbures.

Les résultats obtenus montrent (tableau II) que la productivité en hydrocarbures reste élevée tout au long de la courbe de croissance avec un maximum qui se manifeste vers le milieu de la phase exponentielle. Par ailleurs, on n'observe pas, pour les différents stades de la culture, de modification, ni de la nature, ni des pourcentages relatifs des hydrocarbures produits.

La microscopie électronique a permis de suivre tout au long de la courbe de croissance l'évolution de l'ultra structure des cellules. On constate que celles-ci subissent des transformations très importantes. Dès le milieu de la phase exponentielle on n'observe pratiquement plus de division cellulaire, par contre le volume des hydrocarbures contenus dans la paroi externe augmente largement et le chloroplaste apparaît de plus en plus bourré de grains d'amidon. Une desorganisation progressive du contenu cellulaire se manifeste dès la fin de la phase exponentielle. L'apparition de la phase stationnaire est concomitante à cette desorganisation. Parallèlement l'évolution de la concentration des constituants principaux du milieu de culture et de son pH, en fonction du développement de la culture, a été suivie (tableau III). On note une disparition rapide des phosphates du milieu. Cette disparition, dès le début de la phase exponentielle suggère qu'ils sont absorbés et stockès transitoirement à l'intérieur des cellules. Le fait que leur absence dans le milieu ne constitue pas un facteur limitant la croissance a été vérifié par des expériences utilisant différentes concentrations initiales en phosphates.

TABLEAU II - Productivité pour différents stades de la culture.

Echantillon (1)	Biomasse poids sec g/l	Hydrocarbures g/l	Hydrocarbures % poids sec	Hydrocarbures productivité (2) g/l/j
0	0.21	0.044	21	
1	0.6	0.22	36.5	0.060
2	1.13	0.43	38	0.105
3	2.74	0.96	35	0.265
4	4.5	1.13	25	0.024
5	4.9	1.27	26	0.020
6	4.22	0.8	19	négative

(1) voir sur figure IV les temps de culture auxquels correspondent les prélèvements de ces différents échantillons.
(2) La productivité est calculée pour l'intervalle de temps entre deux prélèvements successifs.

TABLEAU III - Modifications du milieu de culture (batch) en relation avec la courbe de croissance (figure IV).

Echantillon (1)	(Nitrate) mg/l	(Phosphate) mg/l	pH
0	120.5	54	7.6
1	81.6	< 0.5	8.1
2	83.8	< 0.5	7.9
3	83.8	< 0.5	8.6
4	44.8	34.3	7.4
5	33.4	35	7.2
6	63.3	22.3	7.7

(1) voir note 1 tableau II.

Pour les nitrates, on observe par contre une diminution progressive de la concentration qui est encore élevée lorsque s'installe la phase stationnaire. Celle-ci n'est donc pas déclenchée par une starvation totale en cet élément. Néanmoins des essais encore limités semblent montrer que la quantité de biomasse et son taux en hydrocarbures augmentent simultanément lorsqu'on augmente la concentration en nitrates du milieu.

Deux autres problèmes en relation avec la culture de B.b. ont également été abordés, ils se rapportent à l'influence sur la production d'hydrocarbures :

1) de l'origine et de la nature (sauvage ou cultivée) de la souche.
2) de la présence de bactéries accompagnant les cultures.

Les résultats que nous avons jusqu'içi obtenus montrent que suivant la souche de B.b. étudiée, on peut observer des différences importantes non seulement sur la productivité en hydrocarbures, mais aussi sur la nature des hydrocarbures produits. Pour ce qui concerne les bactéries qui accompagnent le plus souvent les cultures de B.b., il est actuellement possible d'affirmer que leur présence n'est pas indispensable pour que l'algue produise des hydrocarbures. Il est également possible de dire que certaines espèces bactériennes qui ont été isolées des cultures se dévelloppent aux dépens des hydrocarbures produits par l'algue et donc diminuent la productivité. Une étude est actuellement en cours, en collaboration avec la Société Elf-Aquitaine, afin de déterminer si d'autres espèces bactériennes exercent au contraire une action favorable à la production d'hydrocarbures.

Les recherches qui viennent d'être brièvement rapportées et qui, pour une part, ont été réalisées dans le cadre de notre contrat CE n° ESE-R-022-F, nous ont permis d'approfondir nos connaissances relatives à la physiologie, au métabolisme et à l'ultra structure de B.b.,ainsi qu'à son comportement en culture. Sur la base des résultats obtenus, dont l'extrapolation permet d'espérer une productivité maximale en hydrocarbures de l'ordre de 50 tonnes par hectare et par an, il est possible de considérer que la production industrielle d'hydrocarbures par culture à grande échelle de B.b. est une chose faisable. Il est maintenant nécessaire de compléter ces recherches par une étude très systématique des paramètres de cultures (suite du contrat CE n° ESE-R-022 F) de façon à aborder dans de bonnes conditions une expérimentation de la culture à l'échelle pilote.

Description du projet.

Les appareils que nous utilisons actuellement pour l'étude des paramètres de cultures (cultures en batch et cultures continues, figures V et VI) sont très éloignés de par leur conception, du cultivateur pilote de 1 m^2 utilisé par l'équipe du CEA pour Porphyridium cruentum, et dans lequel il est envisagé (voir co-projet CEA) de faire, après adaptation, les essais de culture pilote de B.b. C'est pourquoi il paraît indispensable de faire précéder ces essais par une expérimentation en laboratoire,dans un appareil qui sera conçu comme l'élément de base du cultivateur de 1 m^2 et qui comme ce dernier sera équipé d'appareils de contrôle et de régulation automatiques et d'un système de traitement des données.

L'appareil que nous prévoyons, qui pourra fonctionner comme système de culture fermé (batch)ou ouvert (continu), sera composé d'un capteur solaire, constitué par un tube en U, en matériau transparent à la lumière, placé en position horizontale, dans une cuve remplie d'eau. Il sera éclairé par dessus et relié par ses deux extrèmités (entrée et sortie) à une colonne verticale d'aération à laquelle seront raccordés les différents appareils de contrôle. Des orifices seront prévus sur cette colonne pour les entrées d'air (enrichi ou non en CO_2) et de milieu neuf et les sorties de gaz et de culture. Dans un premier temps, nous utiliserons pour l'agitation du milieu et le transport des fluides, les solutions adoptées par l'équipe du CEA pour le cultivateur de 1 m^2. Nous nous proposons de tester par la suite d'autres solutions d'agitation et de circulation, ainsi que différents matériaux transparents pour le capteur solaire. Les appareils de contrôle et de régulation concernent : la mesure du CO^2 fixé en fonction du gradient lumineux, la mesure de l'intensité d'éclairement, la mesure de l'oxygène produit par photosynthèse, le dosage des nutrients (micro et macro) absorbés en cours de culture, la mesure et la régulation du pH et de la température des cultures, l'évaluation de la quantité de biomasse et d'hydrocarbures produits. L'automatisation de ces contrôles permettra de les multiplier et facilitera l'optimisation des conditions de la culture dans des conditions où la lumière est le facteur limitant, notamment : l'optimisation de la concentration des éléments minéraux, de l'apport de CO^2, de la température.

Le problème lié aux contaminations par d'autres microorganismes (maintien d'une culture mono algale - contamination par des bactéries ou champignons prédateurs) lors des prélèvements de biomasse ou des apports de milieu neuf, fera l'objet d'une attention particulière et des solutions devront être trouvées pour en limiter l'occurence.

Enfin les problèmes posés par la récolte de la biomasse et de son traitement pour en séparer les hydrocarbures devront aussi être résolus au niveau de ce micropilote. On peut dès maintenant envisager deux types de procédés :
Pour le premier il y aurait d'abord un traitement mécanique ayant pour but de séparer les cellules des parois contenant les hydrocarbures. Ensuite une décantation permettrait aux hydrocarbures plus légers que les cellules de monter en surface. La récupération des hydrocarbures pourrait alors se faire par filtration de la couche supérieure. Une partie des cellules se trouvant dans la couche inférieure pourrait servir d'inoculum pour de nouvelles cultures.
Pour le deuxième procédé on prévoit de séparer par filtration la biomasse du milieu, puis d'en extraire les hydrocarbures d'abord par pression, ensuite par un traitement aux solvants.
Après séparation des hydrocarbures,une méthanisation de la biomasse résiduelle est,dans tous les cas,envisagée. Elle permettra d'améliorer le bilan énergétique des cultures.
Il sera avant tout tenu compte dans le choix d'une solution aux différents problèmes qui se poseront, de la nécessité de la transposer sans en modifier la conception, à l'échelle des unités de culture de 1 m^2, 10 m^2 et 100 m^2.

L'étude préalable en micro-pilote devrait permettre l'expérimentation des cultures de B.b. dans le cultivateur de 1 m^2, pendant l'exercice 1982-1983, en collaboration avec l'équipe du C.E.A.; nous bénéficierons pour la mise en route de ces essais, et postérieurement pour ceux qui sont prévus à l'échelle des 10 m^2, de l'expérience qui aura été acquise par cette équipe pour la culture de Porphyridium.

<u>Conclusion.</u>

Le projet de micro-pilote de laboratoire que nous proposons pour des essais de culture de B.b. ne doit pas être vu isolément, mais associé au projet de "production et utilisation des algues" présenté par l'équipe du CEA (Cadarache).
Les installations qui sont prévues par cette équipe et qui seront utilisées dans un premier temps pour la culture de Porphyridium cruentum, seront adaptées ensuite à la culture de B.b. Les informations que nous obtiendrons grâce à la mise en oeuvre d'un micro-pilote de laboratoire, dont la construction est projetée, devraient permettre de réussir cette adaptation dans les meilleures conditions. L'expérimentation des cultures qui par ailleurs sera faite dans ce micro-pilote devrait par ailleurs fournir des échantillons d'hydrocarbures en quantité suffisante pour qu'il soit possible de tester ces hydrocarbures sur le plan de leurs utilisations industrielles possibles. Les essais qui seront postérieurement effectués dans les cultivateurs de 1 m^2 et puis de 10 m^2 apporteront les éléments nécessaires à une évaluation relativement précise du prix de revient de ces "hydrocarbures solaires". Actuellement les coûts prévus pour l'obtention d'une biomasse solaire aquatique s'établissent entre 6 F. le Kg (sans séchage) et 8 F. le Kg (avec séchage). Ces prix sont trop élevés pour que l'on puisse envisager de rentabiliser la culture de B.b. pour la produc-

tion d'un "fuel solaire" destiné à se substituer à un "fuel fossile" pour des usages énergétiques. Cependant les hydrocarbures de B.b. présentent sur le plan chimique certaines particularités (longueur de chaîne, insaturation) qui permettent dès maintenant de penser qu'ils pourraient se substituer à certaines matières premières actuellement utilisées,mais en voie de disparition,pour la fabrication de produits de petit tonnage, mais à haute valeur ajoutée.

$$CH_3-(CH_2)_7-CH=CH(CH_2)_n-CH=CH_2$$

avec n = 13, 15, 17

Hydrocarbures prédominants dans les souches cultivées en laboratoire (Origine des souches : Algothèques de Cambridge, Thonon, Göttingen, Austin).

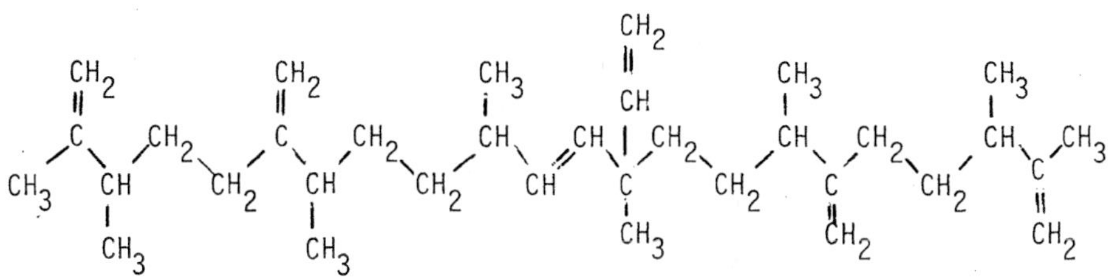

Botryococcène - Hydrocarbure prédominant dans une souche sauvage d'origine australienne.

FIGURE I - Formules des hydrocarbures produits par Botryococcus braunii.

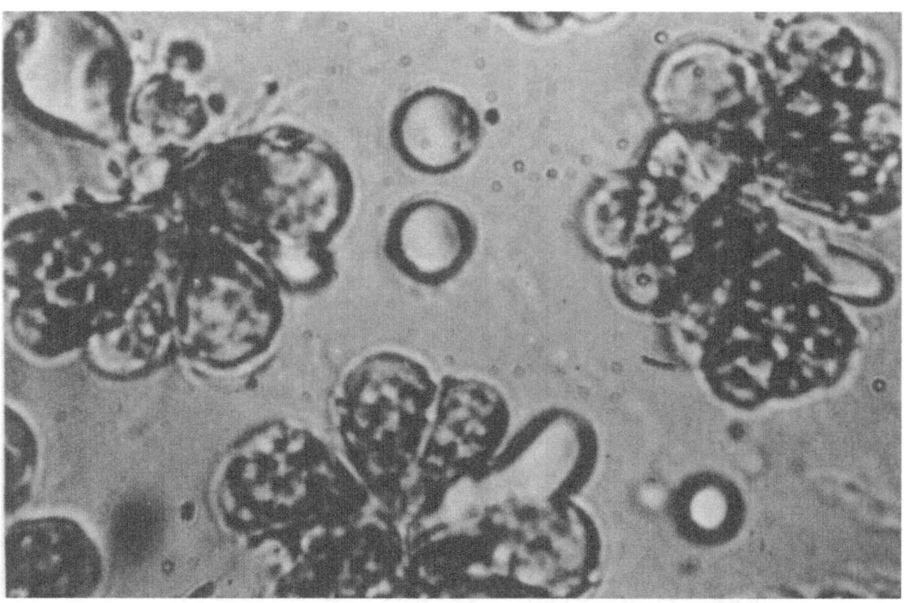

Figure II - Colonies de Botryococcus vues en microscopie optique (x 100) avec leurs globules d'hydrocarbures.

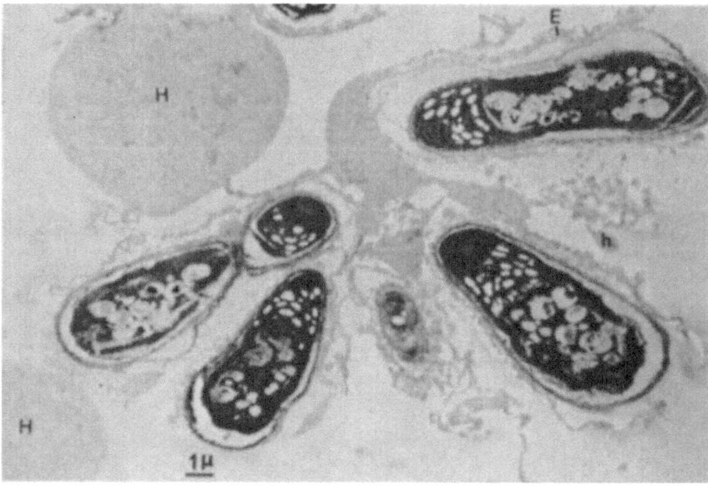

Figure III - Colonie de Botryococcus vue en microscope électronique. H et h = hydrocarbures ; v = vacuoles ; E = paroi externe ; P = chloroplaste avec grains d'amidon.

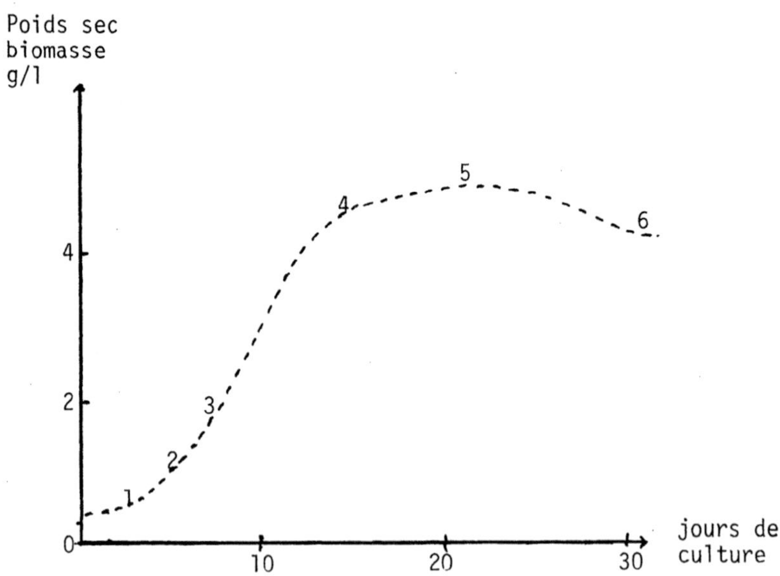

Figure IV - Courbe de croissance de Botryococcus (culture en batch). Les chiffres 1 à 6 situent les prélèvements dont les analyses figurent dans les tableaux II et III.

FIGURE V - Appareillage pour culture en Batch.

FIGURE VI - Appareillage pour cultures continues.

CULTURE DE L'ALGUE BOTRYOCOCCUS BRAUNII A L'ECHELLE PILOTE

Auteurs : M. DESTORDEUR, M.E. ROSSI, C. SIRONVAL

Contrat n° : ESE/P/008/B

Durée : 24 mois (1.7.81 au 30.6.83)

Budget total : 7.864.000 FB

Chefs du projet : R. NOËL, INIEX
C. SIRONVAL, Université de Liège

Organismes contracteurs : INIEX
Laboratoire de photobiologie, Université de Liège

Adresses : INIEX
Rue du Chéra, 200
B - 4000 LIEGE

Laboratoire de photobiologie
Université de Liège, Institut de Botanique B 22
B - 4000 LIEGE (Sart Tilman)

Summary

Two Botryococcus braunii culture experiences have been realised in open air conditions : one in an installation initially designed for Scenedesmus and Chlorella cultures, and the other in a culture installation specially prepared for Botryococcus, taking into consideration the outcome of the first experience. The first mass culture experience has shown that this installation was not adapted to the colonial behaviour of this alga. This experience led us to realise a culture installation of a different type. The mass culture experience in this second installation is still in progress. The results after 7 weeks of culture show Botryococcus to be capable of growing in the open air, at ambient temperature. Scenedesmus and Chlorella appeared in the culture one month after the beginning of the experience, following a relatively important period of rain. The hydrocarbons content of our cultures has shown itself to be feeble : the total hydrocarbons produce is less than 10 % of the dry weight of algae. Future experiences will concern the research of factors furthering the synthesis of hydrocarbons by Botryococcus, as well as the research of a means to eliminate the algae which contaminate the culture medium. These means are described in the following report.

Résumé

Deux expériences de culture de l'algue Botryococcus braunii ont été réalisées en plein air : l'une dans une installation initialement conçue pour la culture de Scenedesmus et Chlorella, l'autre dans une installation de culture mise au point pour Botryococcus tenant compte des résultats de la première expérience. La première expérience de culture en masse a montré que l'installation de culture n'était pas adaptée au comportement colonial de Botryococcus. Cette expérience nous a conduits à mettre au point une installation de culture d'un type différent. L'expérience de culture en masse dans cette seconde installation est toujours en cours. Les résultats obtenus après 7 semaines de culture montrent que Botryococcus est capable de croître en plein air, à la température ambiante. Des Scenedesmus et des Chlorella sont apparus dans la culture 1 mois après le début de l'expérience, suite à une période de pluie relativement importante. La teneur en hydrocarbures de nos cultures s'est avérée faible (rendements en hydrocarbures inférieurs à 10 % du poids sec d'algues). Les expériences futures concerneront la recherche de facteurs promouvant la synthèse des hydrocarbures par Botryococcus, ainsi que la recherche d'un moyen d'élimination des algues contaminant le milieu de culture. Ces moyens sont décrits dans le rapport qui suit.

1.1. Introduction

Jusqu'à présent, les recherches entreprises sur Botryococcus braunii ont presque exclusivement concerné les milieux de culture appropriés à l'algue, ainsi que l'étude chimique des hydrocarbures produits par l'algue cultivée en laboratoire, ou prélevée de son milieu naturel. Bien que la plus forte prolifération de l'algue et sa plus haute teneur en hydrocarbures aient été observées dans le cas d'algues prélevées dans la nature, on n'a encore jamais essayé de la cultiver en plein air et en masse. On se propose de cultiver Botryococcus en masse en plein air dans le but de produire des quantités appréciables d'hydrocarbures (par unité de surface et de temps).

1.2. Description des installations de culture et des méthodes de mesure

A. Le dispositif de culture d'algues à l'échelle pilote installé à l'Institut de Botanique est représenté par la figure I. Ce dispositif de culture massive fonctionne en continu à l'air libre et à température ambiante. Les algues en suspension dans le milieu de culture s'écoulent en mouvement turbulent sur un plan incliné d'une surface de 32 m^2 et d'une pente de 4 %. Le système est muni d'une pompe qui permet une circulation continue des algues, leur contact avec l'air atmosphérique et l'action de la lumière solaire.

B. Le dispositif de culture d'algues à l'échelle pilote installé à l'INIEX est représenté à la figure II. La suspension d'algues est placée dans une cuve en acier inoxydable (dimensions : 1 x 1 x 0,75 m). Le fond de la cuve est muni, dans sa partie la plus basse, d'une vanne permettant de vider facilement la cuve de son contenu. Un mélangeur, dont l'hélice tourne à une vitesse de 700 tours/min, est adapté à la cuve. Le jour, le mélangeur brasse la culture afin de présenter toutes les algues à la

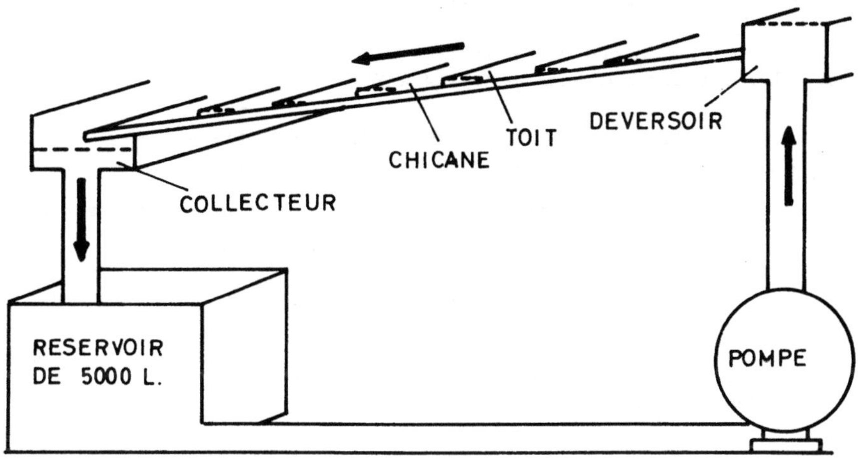

Fig. I : Dispositif de culture d'algues installé à l'Université de Liège

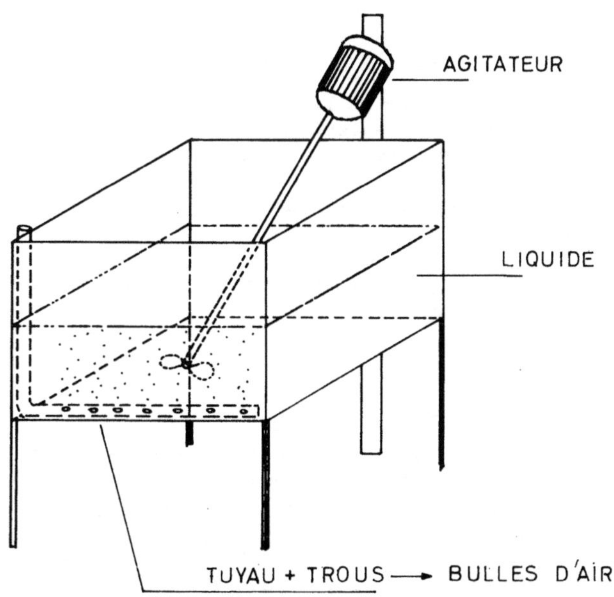

Fig. II : Dispositif de culture d'algues installé à l'INIEX conçu pour la culture de Botryococcus

lumière. La nuit, une minuterie arrête l'agitation. La culture est aérée 24 h sur 24 (débit d'air : 66 litres/min). Une moustiquaire placée sur la cuve empêche les retombées de feuilles, d'insectes, etc.. dans la cuve.

C. La souche de Botryococcus braunii utilisée provient de la Cambridge culture collection (n° LB 807/1).

La composition du milieu Chu modifiée est la suivante (g/litre) :

KNO_3 : 0,2 K_2HPO_4 : 0,02 $MgSO_4.7H_2O$: 0,1

$CaCl_2.6H_2O$: 0,08 Fe citrate : 0,02 Ac.citrique : 0,1

Le pH est ajusté à 7 avec une solution de KOH.

La croissance des algues est évaluée par comptages à la cellule de Thoma.

Les algues et les hydrocarbures qui y sont associés sont récoltés par centrifugation.

La récolte est lyophillisée afin de permettre le calcul de rendements en hydrocarbures précis se rapportant au poids sec des algues. L'extraction des hydrocarbures externes aux cellules se fait à l'hexane; celle des hydrocarbures internes au chloroforme-méthanol (1 : 2). La purification des extraits se fait par chromatographie sur colonne d'alumine.

L'analyse quantitative des hydrocarbures se fait par pesée et par chromatographie en phase gazeuse (GLC); l'analyse qualitative par GLC et par spectroscopie infrarouge (IR).

GLC : L'appareil utilisé est un Perkin Elmer de type F 22. Le détecteur est à ionisation de flamme. La colonne employée est une colonne capillaire.

IR : L'appareil utilisé est un Perkin Elmer modèle 21. Pour l'enregistrement des spectres, les hydrocarbures sont placés en sandwich entre deux pastilles de KBr.

Les nitrates du milieu de culture sont dosés selon la méthode au salicylate de sodium (1).

Les phosphates du milieu de culture sont dosés selon la méthode au molybdate d'ammonium (2).

1.3. Résultats

A. Cultures semi-stériles

Des cultures semi-stériles de Botryococcus d'un volume de 1,5 à 10 litres ont été réalisées pour déterminer les conditions du milieu qui favorisent la croissance de l'algue et sa production en hydrocarbures, ainsi que pour inoculer les cultures de plein air à l'échelle pilote. Les conditions de culture testées sont celles qui sont maîtrisables dans le cas de la culture en plein air, à savoir :

l'agitation,
l'aération,
la composition du milieu,
le pH.

a) Résultats concernant la croissance des algues

Toutes autres conditions étant identiques, les cultures agitées présentent, par rapport aux cultures non agitées, une croissance plus rapide.

De même, les cultures aérées présentent, par rapport aux cultures non aérées, une croissance plus rapide.

La croissance la plus forte est obtenue dans le cas de cultures à la fois agitées et aérées.

Les milieux de culture suivants ont été testés :

. milieu Chu modifié (composition, voir chap. 1.2.)
. eau de mer
. eau de pluie
. eau de distribution

Ces milieux ont été choisis pour les raisons suivantes : le milieu Chu modifié parce qu'il a été spécialement conçu pour la culture de Botryococcus en laboratoire ; l'eau de mer et l'eau de pluie parce que l'on a mentionné l'existence de Botryococcus dans de tels milieux à l'état naturel (3, 4) ; l'eau de distribution parce que l'on y dissout les sels pour constituer le milieu de culture des algues cultivées en masse.

Toutes autres conditions étant identiques, le nombre de Botryococcus/ml obtenu après 4 semaines de culture décroît pour les milieux suivants : milieu Chu modifié, eau de pluie, eau de distribution, eau de mer.

L'influence du pH sur la croissance de l'algue a été étudiée de 2 manières :

- en inoculant Botryococcus dans du milieu Chu modifié, dont le pH est ajusté à 4, 7, 9 au début de l'expérience par des solutions de HCl ou KOH ;
- en inoculant Botryococcus dans du milieu Chu modifié, dont le pH est ajusté à 4, 7, 9 au départ de l'expérience et réajusté à ces valeurs respectives toutes les semaines par des solutions de HCl ou KOH.

L'expérience montre que Botryococcus croît de la même façon aux différents pH. L'ajustement du pH toutes les semaines ne modifie pas la croissance de l'algue.

b) Résultats concernant la teneur en hydrocarbures des cultures

Dans les conditions de culture testées, les cultures ont une teneur faible en hydrocarbures totaux (hydrocarbures externes + hydrocarbures internes). Les rendements en hydrocarbures totaux sont inférieurs à 10 % par rapport au poids sec d'algues.

Les rendements en hydrocarbures externes sont de l'ordre de 10 fois supérieurs à ceux des hydrocarbures internes.

c) Résultats concernant l'analyse qualitative des hydrocarbures

L'absence de bandes d'absorption caractéristiques de groupements étrangers aux liaisons C-C ou C-H, lors de l'enregistrement des spectres IR des hydrocarbures (externes et internes), prouve que l'on obtient des extraits hydrocarbonés purs. La présence de bandes d'absorption caractéristiques du groupe vinyl (= CH_2) indique l'existence d'oléfines dans nos extraits (fig. III).

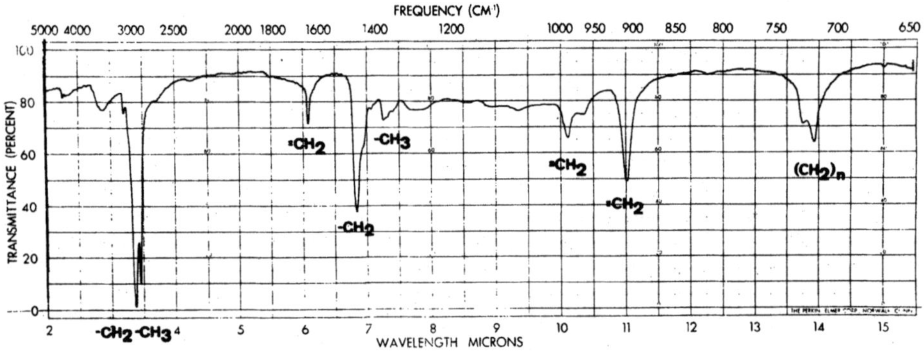

Fig. III : Spectre IR des hydrocarbures extraits de Botryococcus

La GCC des extraits hydrocarbonés (externes et internes) montre qu'il s'agit d'un mélange complexe d'hydrocarbures (fig. IV). Les principaux d'entre eux ont un nombre d'atomes de carbone égal à 27-29-31. Les hydrocarbures internes présentent parfois, en plus des hydrocarbures communs aux hydrocarbures externes, des hydrocarbures plus courts (C_{15} - C_{17}). On n'a pas noté de corrélation entre telle ou telle condition de culture et la présence de tel ou tel hydrocarbure.

B. Cultures en masse (non stériles)

Tenant compte des résultats obtenus lors des expériences de culture semi-stériles, Botryococcus a été cultivé dans une première installation pilote.

1ère installation pilote (Université de Liège - fig. I, chap. 1.2.)

L'installation pilote fonctionnant à l'Université de Liège est initialement conçue pour la culture de Scenedesmus et Chlorella, algues ayant besoin d'un milieu bien agité et bien aéré pour atteindre leur vitesse de croissance maximale.

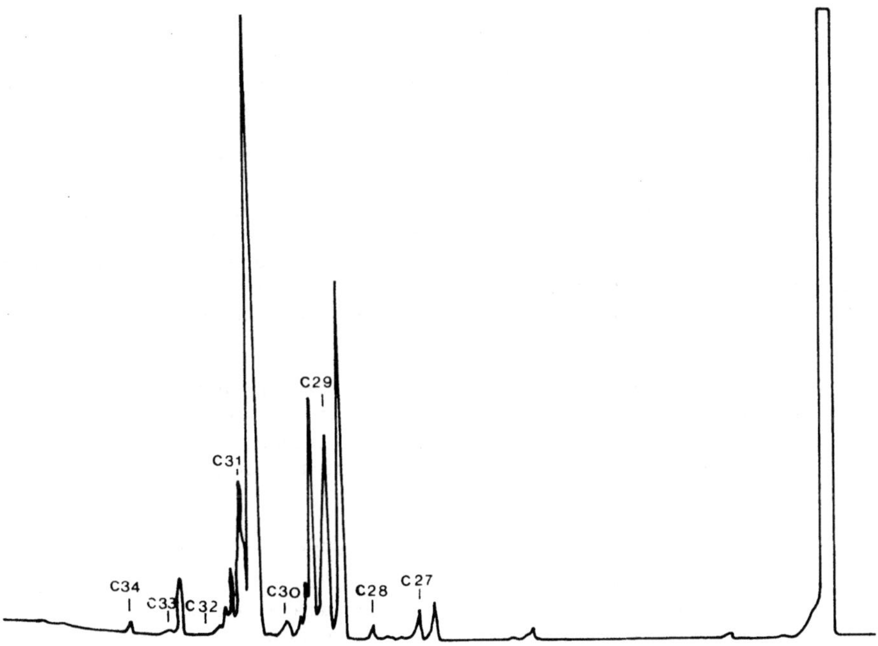

Fig. IV : Chromatogramme des hydrocarbures extraits de Botryococcus

Des cultures semi-stériles de Botryococcus ont été préparées pour inoculer au mois de mai l'une des 4 cuves du pilote ; les 3 autres cuves voisines de la première servant à la culture de Scenedesmus et Chlorella durant l'expérience. Le milieu de culture utilisé est le milieu Chu modifié, dont le pH est ajusté à 7 au départ.

Les <u>résultats</u> de l'expérience sont les suivants :

- La concentration en Botryococcus augmente la première semaine puis se stabilise pendant six semaines.

- La proportion de cellules de Botryococcus isolées par rapport aux cellules de Botryococcus groupées en colonies augmente au cours du temps.

- Scenedesmus et Chlorella contaminent la culture une semaine après l'inoculation. Peu de temps après leur apparition dans le milieu, les rapports $\frac{\text{nombre de Botryococcus}}{\text{nombre de Scenedesmus}}$ et $\frac{\text{nombre de Botryococcus}}{\text{nombre de Chlorella}}$ diminuent rapidement au cours du temps, indiquant que les vitesses de croissance de Scenedesmus et de Chlorella sont plus rapides que celle de Botryococcus.

- La teneur en hydrocarbures d'une telle culture est faible : les rendements en hydrocarbures totaux sont inférieurs à 10 % par rapport au poids sec des algues.

La croissance de Botryococcus plus lente que celle de Scenedesmus et de Chlorella, qui poussaient dans les autres cuves et envahissaient la culture de Botryococcus, la mauvaise adaptation du système d'agitation de la culture au comportement colonial de l'algue ont déterminé l'arrêt de l'expérience et la mise au point d'une autre installation pilote.

2ème installation pilote (INIEX - fig. II, chap. 1.2.)

Cette seconde installation est placée sur l'un des toits de l'INIEX, Institut éloigné des cultures de Scenedesmus et Chlorella précédemment mentionnées. L'agitation du milieu est produite non plus par écoulement violent de la culture le long d'un plan incliné et refoulement de la suspension algaire par une pompe, mais par un mélangeur qui brasse le milieu de culture.

Des Botryococcus ont été ensemencés dans cette cuve le 18/8/81. Le milieu de culture est le milieu modifié Chu. Le pH était de 8,5 au départ de l'expérience. La croissance des algues, la teneur en NO_3^{2-}, PO_4^{3-} du milieu, la pluviosité et la température sont évaluées chaque semaine.

Les résultats au 7/10/81, soit après 7 semaines de culture, sont les suivants :

- Pendant 6 semaines, on observe une augmentation du nombre de Botryococcus dans le milieu (fig. V). On ne note pas d'augmentation de la proportion de cellules isolées par rapport au nombre de colonies de Botryococcus.

- Après 3 semaines, on observe l'apparition de Scenedesmus puis de Chlorella dans le milieu. Alors que dans le cas de la première expérience de culture pilote, Scenedesmus et Chlorella auraient pu être amenés dans le milieu suite à un nettoyage imparfait de l'installation précédemment utilisée pour la culture de ces espèces, dans le cas présent ces algues ne peuvent provenir que d'une contamination extérieure au dispositif de culture. L'apparition de ces algues coïncide, par ailleurs, avec une période de pluviosité relativement importante.

- Après 6 semaines, on observe une diminution du nombre d'algues en général pouvant être corrélée à une diminution de la température moyenne du milieu.

- L'évolution de la concentration en nitrates observée dans la culture est semblable à celle rencontrée chez des cultures unialgales de Botryococcus, Scenedesmus ou Chlorella (fig. VI). Par contre, l'évolution de la teneur en phosphates dans le milieu est particulière à Botryococcus (fig. VII). En effet, des cultures unialgales de Botryococcus présentent une forte diminution de la teneur en phosphates du milieu dès le premier jour qui suit l'inoculation. Dans de telles cultures, on observe cependant la présence de phosphates 12 semaines après l'inoculation, alors que dans le cas de la culture de plein air, cette teneur est nulle 5 semaines après le début de l'expérience. Dans le cas de la culture, Scenedesmus et Chlorella se sont développés, alors que la teneur en nitrates et en phosphates était faible. La présence de ces contaminants a contribué à épuiser le milieu. Outre la corrélation entre la diminution du nombre d'algues dans la culture et la température moyenne, on peut également attribuer cette diminution à l'épuisement du milieu en ces éléments.

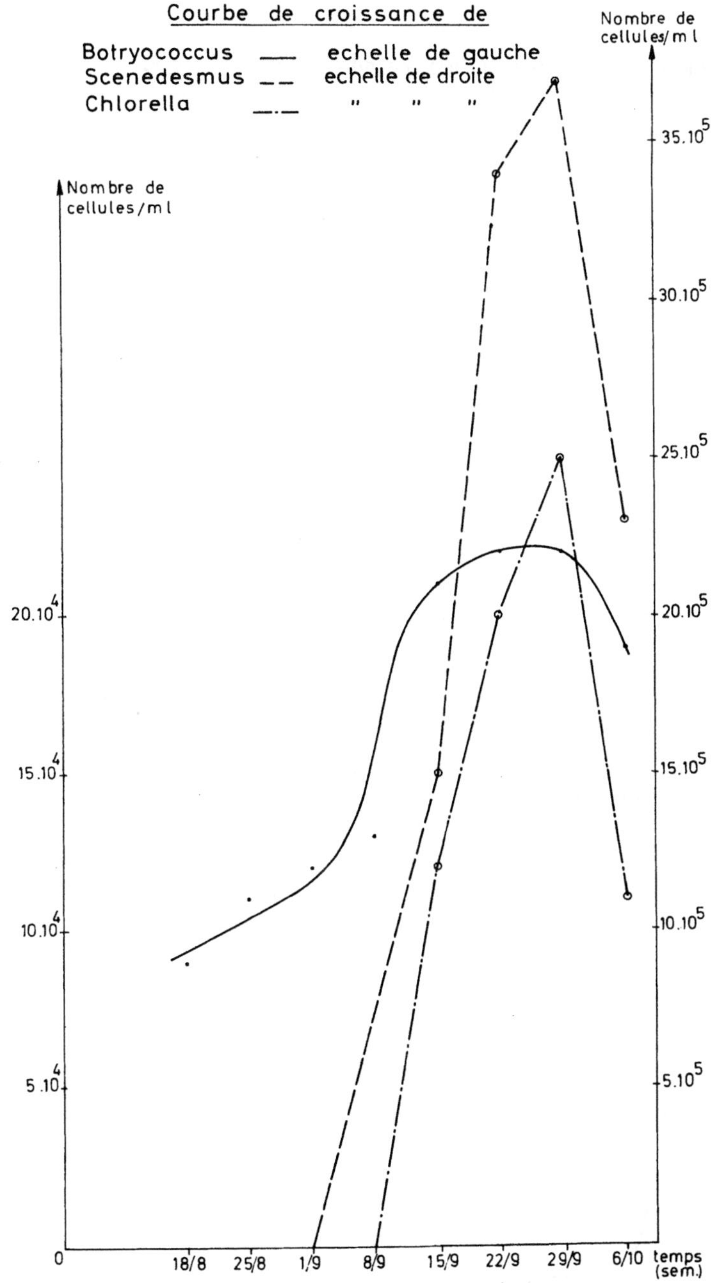

Fig. V : Courbes de croissance de Botryococcus, Scenedesmus, Chlorella
(comptages à la cellule de Thoma)

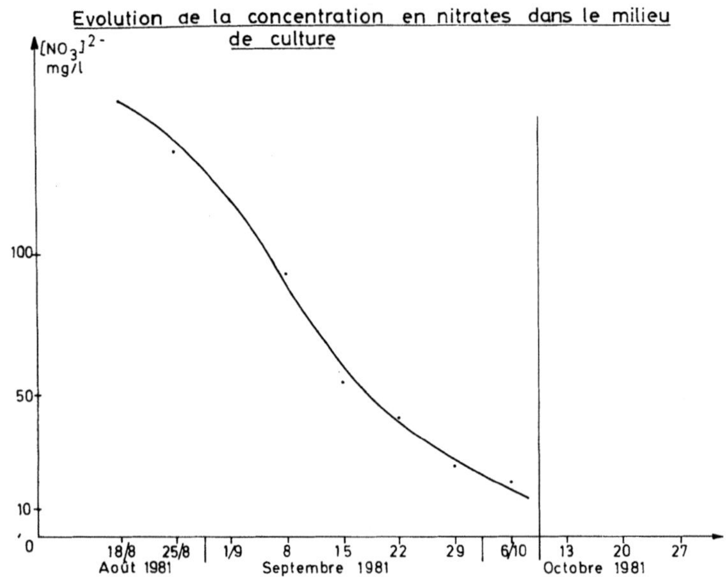

Fig. VI : Evolution de la concentration en nitrates dans le milieu de culture (dosage colorimétrique)

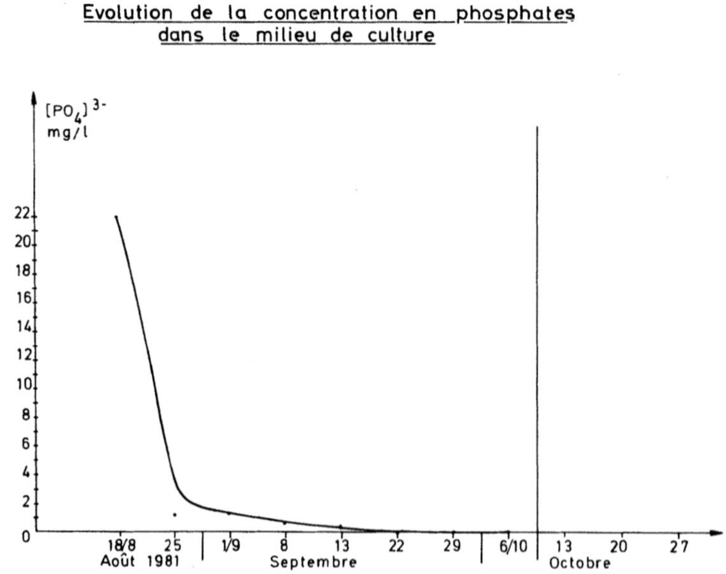

Fig. VII : Evolution de la concentration en phosphates dans le milieu de culture (dosage colorimétrique)

- La teneur en hydrocarbures de la culture est faible : les rendements en hydrocarbures totaux sont inférieurs à 10 % par rapport au poids sec d'algues.

1.4. Analyse des résultats et commentaires

Les expériences de culture en masse montrent que les problèmes majeurs auxquels on est confronté sont l'élimination des algues contaminant le milieu et la recherche de facteurs promouvant la synthèse des hydrocarbures par Botryococcus.

La contamination d'une culture par d'autres organismes est inévitable dans le cas de culture pilote en plein air, les spores d'algues étant véhiculées par le vent et l'eau de pluie. Dans nos expériences de culture pilote, deux espèces d'algues seulement entrent en compétition avec Botryococcus : Scenedesmus et Chlorella. Il s'agit donc de trouver le moyen d'éliminer efficacement ces algues ou d'empêcher leur croissance. Ce moyen peut être de nature :
- biologique
- nutritionnelle
- chimique
- physique

a. Biologique

On a constaté que quand on cultive en masse, en plein air, l'algue filamenteuse Hydrodictyon dans des bassins alimentés par de l'eau de rivière, la culture se contamine par l'algue filamenteuse Chladophora, par des Lemma, etc..., mais on n'observe pas de prolifération de Scenedesmus et Chlorella. Des expériences consistant à introduire, dès le départ, dans une culture en masse de Botryococcus, une algue filamenteuse telle qu'Hydrodictyon, seront réalisées afin de voir si cette algue empêche la prolifération de Scenedesmus et de Chlorella. Les algues filamenteuses étant morphologiquement très différentes de Botryococcus, elles seraient par conséquent facilement séparables par filtrage.

b. Nutritionnelle

Actuellement, on ne sait pas encore combien de temps Botryococcus et les algues contaminant la culture vont pouvoir subsister dans un milieu appauvri. La comparaison systématique des besoins nutritionnels de Botryococcus, d'une part, de Scenedesmus et Chlorella, d'autre part, est à faire pour déterminer les concentrations limitantes en éléments nutritifs qui sont supportées par Botryococcus, mais non par Scenedesmus et Chlorella.

c. Chimique

L'introduction dans le milieu de culture d'une molécule (organique ou non) à une concentration supportée par Botryococcus, mais létale pour Scenedesmus et Chlorella, est à envisager en cas d'échec des deux moyens précédemment cités.

d. Physique

La taille des Botryococcus, d'une part, des Scenedesmus et Chlorella, d'autre part, ainsi que leurs densités différentes, devraient permettre de les séparer par un moyen physique (par exemple, l'écrémage). Ce moyen doit être peu coûteux et permettre la récolte des Botryococcus et des hydrocarbures qui leur sont associés.

Il faut remarquer que les moyens qui seront utilisés pour éliminer les contaminants ne sont pas forcément ceux qui vont favoriser la synthèse d'hydrocarbures par Botryococcus. Afin de connaître l'impact de tel ou tel moyen d'élimination des contaminants sur la production en hydrocarbures de Botryococcus, les hydrocarbures présents dans la culture seront régulièrement dosés.

Afin d'augmenter la teneur en hydrocarbures de nos cultures, on cultivera d'autres souches de Botryococcus réputées receler des quantités d'hydrocarbures plus importantes que la souche que nous avons étudiée jusqu'à présent.

CONCLUSIONS

A l'issue de nos travaux, il s'avère que les méthodes de comptage des algues, de dosage et d'analyse des hydrocarbures, de dosage de certains éléments du milieu sont au point.

Sur base des résultats obtenus lors du premier essai de culture pilote, une installation a été spécialement mise au point pour la culture de Botryococcus. Cette installation est encore imparfaite et de petite taille, mais son fonctionnement est adapté au comportement colonial de l'algue.

Les essais de culture en plein air ont principalement montré que :

1) Botryococcus est capable de croître en plein air à la température ambiante.
2) Dans les conditions climatiques belges, deux espèces d'algues seulement concurrencent Botryococcus dans son milieu : Scenedesmus et Chlorella.
3) La teneur en hydrocarbures des cultures est faible : les rendements en hydrocarbures totaux sont inférieurs à 10 % par rapport au poids sec des algues.

Il ressort donc que les questions suivantes doivent être résolues en priorité :

1) Comment augmenter la teneur en hydrocarbures des cultures ?
2) Comment empêcher Scenedesmus et Chlorella de se développer dans les cultures de Botryococcus ?

Dans l'immédiat, on se propose, comme prévu dans le projet initial, de continuer à décrire l'évolution de la culture pilote en n'apportant pas volontairement de modification du milieu. Cette description permettra notamment de savoir si Botryococcus est plus résistant que Scenedesmus et Chlorella face à l'épuisement du milieu et à la baisse de température qui va s'opérer dans les prochaines semaines.

D'autre part, une seconde cuve semblable à la première sera construite de façon à pouvoir disposer d'une cuve témoin et d'une cuve dans laquelle on va tester l'influence de différents paramètres (culture d'une autre souche de Botryococcus, introduction d'algues filamenteuses, modification de la concentration en tel ou tel élément nutritif) sur la croissance des algues et la production d'hydrocarbures par Botryococcus.

REFERENCES

1. RODIER : Analyse de l'eau
Ed. Dunod, 5ème édition, p 170.

2. EMPAIN : Communication personnelle.

3. GORHAM : Limnology and Oceanography, vol 2 (1957),p 22.

4. HILLEN-WAKE : "Solar Oil" - Liquid Hydrocarbon Fuels from Solar Energy via Algae -
AIE National Conference, Newcastle, feb. 1979.

FUEL GAS PRODUCTION BY MARICULTURE ON LAND

Author : K. Wagener

Contract number : ESE - P - 021 - D

Duration : 24 months

Total budget (requested): DM 862 360,--; CEC contribution: 50 % (expected)

Head of project : Prof. Dr. K. Wagener

Contractor : Technical University Aachen
Templergraben 55
D-5100 Aachen (FRG)

Subcontractor : CSMA - Istituto di Microbiologia Agraria e Tecnica dell' Universitá di Firenze

Summary

The current stage of the "Mariculture on Land" project after 3 1/2 years of activity seems to justify an upscaling to several thousand m^2 of pond area. The average productivity over 36 months in Lamezia/ Calabria is 65 t dry matter/ha/yr. The methane yield from this harvest is presently 13 000 m^3/ha/yr. A cheap way of pond construction has been developed which means a capital investment of 10,-- DM per m^2. Also a mixing procedure was developed which does not consume more than 4 % of the energy stored in the grown biomass. Mineral recycling in form of the sludge directly given back into the algaeponds has successfully been practiced. Starting from this basis the proposed upscaling would give the opportunity to test the behaviour of the system at the pilot scale level. The necessary facilities for this are already partially installed, but for the construction of 1000 or 2000 m^2 additional pond area money is needed for materials and additional personnel. The installation of a 1 m^3 fermenter is already part of the present contract. A cost analysis based on the presently obtained data shows that biofuel from algae is well within the range of methanol from wood and ethanol from sugar cane.

1. Introduction

The general goal of the project is the development and testing of a system for intensive energy farming without pressure on land, based on seawater and using arid coastal areas.

Actual work was conducted and is going on along the following lines:

- Growing microalgae for biomass (because of their exceptional high productivity) - but the system is capable of growing also any other marine phytoplancton
- Converting the harvest via anaerobic fermentation to methane - but any other route is open for testing (as e.g. low temperature catalytic conversion to oil or fermentation of polysaccharides to ethanol).
- Covering the energy demand for seawater pumping from the systems output - but this could also be taken from other sources (like solar radiation, wind, wave power etc.)

Thus the present project is a special version of the general concept of Mariculture on Land; see Figure 1.

2. The current state of the MC1 project

as developed through the contracts 462 - 78 - 1 ESD
and ESE - R - 009 - D

2.1 Algal strains

Biological screening of various microalgae showed several suitable strains. Presently the management of the big ponds is adapted to the properties of Tetraselmis, a green microalga (1) (2).

2.2 Biomass yield

Outdoor cultures in Lamezia/Calabria continuously in operation since 1978, have a mean productivity of 65 t dry organic matter (D.O.M.)/ha/yr (average over 12 months, 3 years) (1)(2).

This means 2.0 % fixation of the global radiation (as measured at the Messina station) in the form of ash-free volatile organic matter (V.O.M.).

During the summer period (May - October) the yield is 91 t/ha/yr $\hat{=}$ 2.8 % fixation in V.O.M..

2.3 Fuel yield

Yield of anaerobic fermentation: 200 l Methane/kg D.O.M. = 280 l/kg V.O.M. = 13 000 m^3/ha/yr $\hat{=}$ 0.8 % of global radiation fixed as methane.

For comparison:

	t/ha/yr	solar radiation kWh/m^2/yr	Fuel yield	% of global rad. stored in fuel
Short Rot. Forestry in EC countries (3)	9	1200	3.0 t Methanol	0.16 %
Sugar cane (Brasil)	25	2600	4.4 t Ethanol	0.12 %
MC1 algae (Italy)	65	1750	13 t Methanol	0.44 %

2.4 The ponds

In operation:

about 12 experimental ponds for screening (3 m² each):	40 m²
4 ponds for production, 20 by 2 m each:	160 m²
(see Figures 2, 3, 4)	
Total	200 m²

Partially constructed

Larger ponds with a total area of (in operation in April 1982)	700 m²

Future ponds (see Figure 5) will cost in German 1981 prices:

	DM/m²
Soil movement, planishing	2.00
Pond walls	1.30
Bottom (compacted clay)	5.00
Finishing	0.60
Tubing for harvesting	1.00
Total costs of pond construction	10.00 DM/m²

2.5 Mixing in algae ponds

by low speed board mixing (see Figure 6):

capital investment	1.50 DM/m²
energy consumption	0.1 W_{mech}/m²
	≙ 4 % of stored energy

2.6 Costs of MC1 Biofuel (1981)

		TDM/ha
Capital investment:	Pond	100
	Mixer	15
	Seawater pumping	25
	Fermenter	15
	Storage for biogas	3
	Total	158
	10 % return on invest. p. a.	16 TDM/ha/yr
Operation costs:	Repair (= 3 % of C.I.)	5 TDM/ha/yr
	Labor* (0.5 person x 1 yr)	6 TDM/ha/yr
	Total	11 TDM/ha/yr

(* = southern European income level in rural areas)

Total costs p.a. 27 TDM/ha/yr

From this follows (1 US Dollar = 2.3 DM) for MC1 methane:

23 \$/GJ (EC price and productivity level)
10 \$/GJ (southern country price and prod. level)

A future decrease is mainly expected from larger units.

For comparison: 9.1 \$/GJ Gasoline (tax free, 1981)
9 \$/GJ Methanol from energy forest (4)
18 \$/GJ Ethanol from sugar cane (Brasil)

3. Proposal of a pilot project on MCl Energy Farming

3.1 Justification

The results obtained during the last 3 1/2 years and as summarized in the previous chapter, seem to justify an upscaling to some 1000 m^2 of pond area. The pond management as it has successfully been practiced with ponds of 40 m^2, may not necessarily be proper for larger units, but needs perhaps some alterations. It must be stressed that besides our effords no larger algae ponds have been run exclusively for energy farming. Therefore new aspects of energy saving in all operations (like mixing and harvesting) have to be taken into account. Sewage and waste water ponds have different objectives. On the other hand, it seems very likely that algae will find their place in the future energy market, and this for several reasons: highest productivity = shortest doubling time, easy fermentation, and - last not least - they need no fertile land to grow. It is therefore a challanging task to master the remaining obstacles on this way, and that means mainly to reduce the capital investment and to practice algae farming preferably in warm climates for higher yields.

3.2 Planned actions for the pilot phase: constructions

- Finishing the already pre-constructed ponds of 20 m length on the already existing platform, giving an additional total of 700 m^2.
- Installing at least 4 big ponds, 4 by 60 m each, giving a total of 960 m^2. They shall be constructed according to the new version of cheap ponds as illustrated in Figure 5.
- Installing in all new ponds the low energy consuming mixing system as developed recently (see Figure 6) which consumes not more than 4 % of the energy stored in the grown biomass.
- Installing a gas stripper: the biogas is passed upwards through a vertical exchange column where the CO_2 is washed out with seawater running down. This measure gives almost pure methane and allows at the same time to recycle this fraction of the primarily fixed carbon.
- Installing a storage for the produced fuel gas: plastic bag placed in a simple shelter.

3.3 Planned actions: operational programme

- Running all large ponds (with a total area of about 1800 m^2) on a routine basis for maximal biomas yield. This will give about 10 - 12 tons dry matter per year.
- Processing most of this harvest through the 1 m^3 fermenter installed on the pond site by the group of Louvain university (Profs. Nyns and Naveau) to biogas.
- Testing on an experimental scale in collaboration with Prof. E. Bayer (Inst. of Organic Chemistry, University of Tübingen) the low temperature catalytic conversion (LTCC) of dried algae at 300°C giving directly liquid hydrocarbons. The aim is to clarify which route has the better yield in terms of productivity and energy budget: LTCC of any phototrophic biomass or the

4. Time table

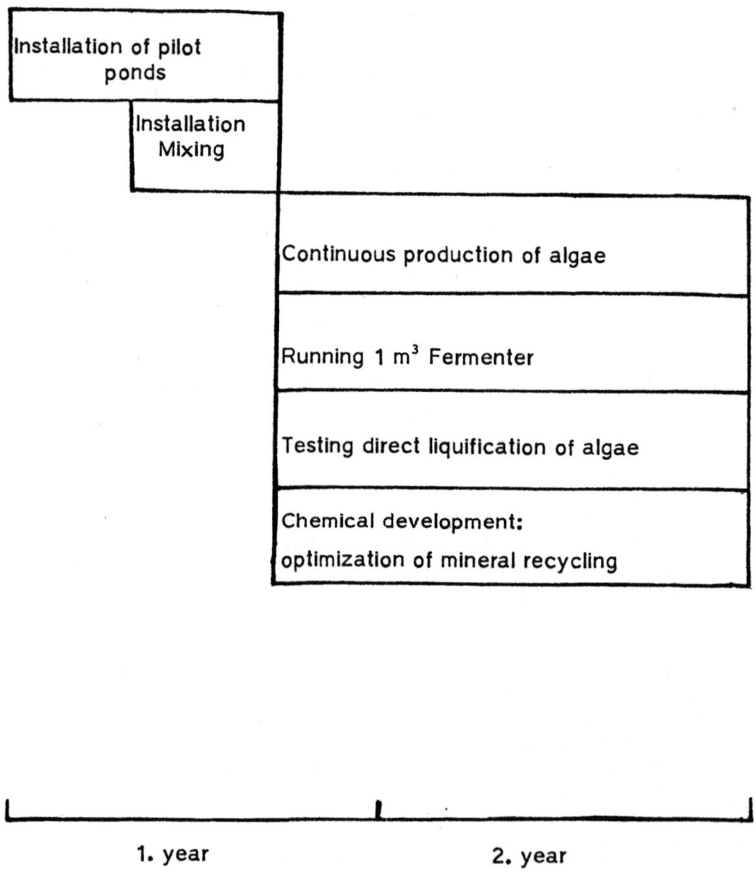
Installation of pilot ponds
Installation Mixing
Continuous production of algae
Running 1 m³ Fermenter
Testing direct liquification of algae
Chemical development:
optimization of mineral recycling
1. year
2. year

extracellularly produced hydrocarbons of Botryococcus braunii.

- Testing various alternations of the pond management for optimizing the whole process
- Practicizing mineral recycling by direct recyclation of the sludge into the algae ponds, as it has successfully been tested in smaller ponds. Only few percents of the daily mineral nutrients requirement has to be added additionally. This means an almost closed mineral cycle.
- Evaluation of all data for systems, energy, and costs analysis.

5. Breakdown of the cost of the programme

5.1 Summary table

Staff expenses	697 800 DM
Mission expenses	45 760 DM
Other expenses	118 800 DM
Total	862 360 DM

5.2 Comments

- About 65 % of the total budget goes to the Italian subcontractor, and there mainly to pay the persons working at the ponds in Lamezia all the year round.
- The travel expenses result from the fact that the main activities are located in three places (Lamezia/Calabria, Florence, and Aachen/Jülich) with a considerable distance from each other.
- Some additional money (50 % as the Commission's share of additional 100 000 DM) for a very desired extension of the total pond area to 5 000 m^2 would bring the whole plant to a demonstration plant level, and would allow to run a stationary combustion engine (e.g. to produce electricity) permanently.

References

(1) Final report of contract 462 - 78 - 1 ESD (Commission of the European Communities)

(2) Status report on project ESE - R - 009 - D. Presented at the Coordination meeting of contractors, June 1981, Copenhagen.

(3) P. Bouvarel: The outlook for energy forestry in France and in the European Economic Community. In: W. Palz, P. Chartier, D. O. Hall (eds), Energy from Biomass. Applied Science Publishers, London 1981. p. 172.

(4) Cited after W. Palz, P. Chartier (eds): Energy from Biomass. Applied Science Publishers, London 1980; p. 208. Cost of wood input revised after P. Chartier: Prospects for energy from biomass in the European community. In: W. Palz, P. Chartier, D. O. Hall (eds): Energy from Biomass. Applied Science Publishers, London 1981 p. 30, Table 4.

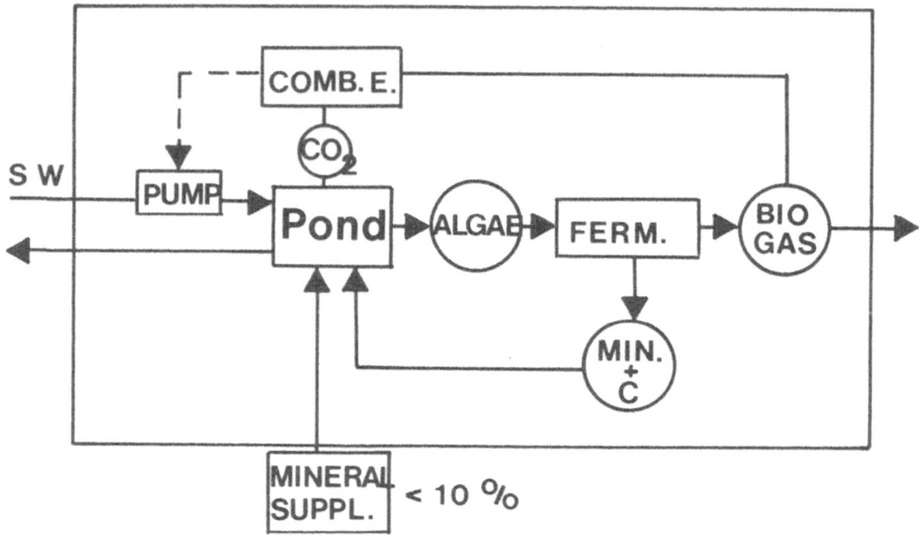

Figure 1: Scheme of the process for the pilot phase

Figure 2: A series of ponds, 2 by 20 m each, at Lamezia/Calabria.

Figure 3 and 4: Ponds at Lamezia. The mixers are floating rafts carrying one mixing board at each end, and are pulled forth and back through the pond. The mixers of two parallel ponds are always operated together.

Fig. 5: **POND CONSTRUCTION**

LEVELING: 2 DM / m^2

CONCRETE BLOCS

9 DM / m of wall

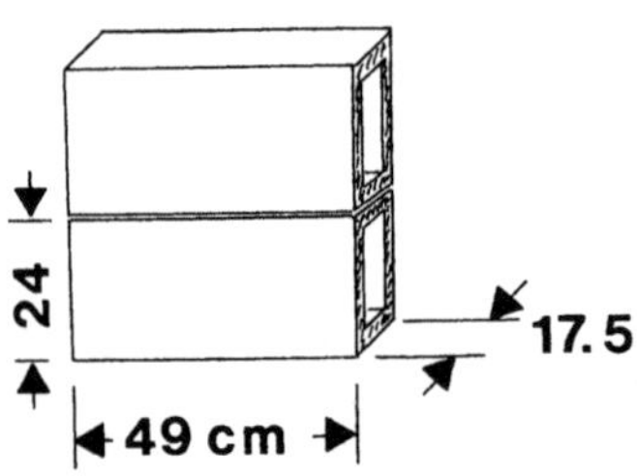

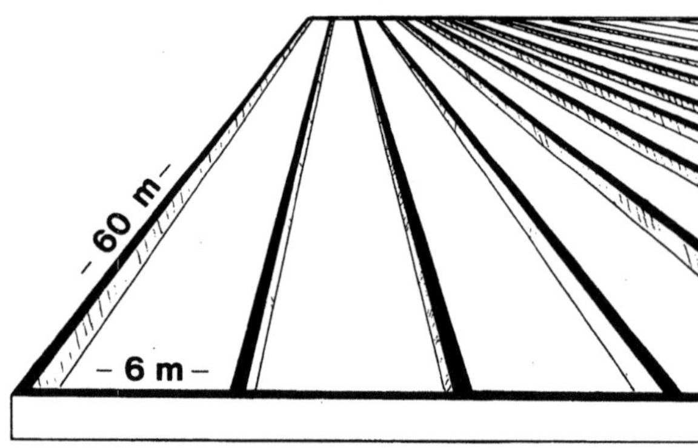

POND WALLS: 1.3 DM / m^2

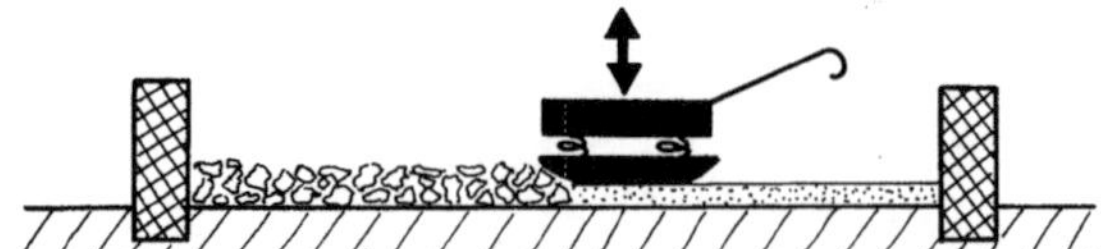

CHIPPED Clay — **4 DM / m^2**

COMPACTING — **1 DM / m^2**

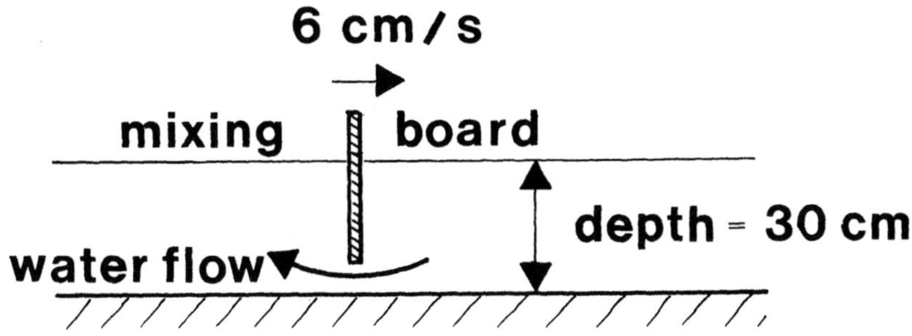

Figure 6: Longitudonal section through a channel-type pond showing the mixing procedure. The mixing board closes the channel leaving open only a slit at the bottom. Moved at low speed, it forces all the water through the slit which blows up very effectively all settled cells.

LIST OF PARTICIPANTS

ALFANI, F.	Ist. di Principi di Ingegneria Chim. Università di Napoli Ple Tecchio I - 80125 NAPOLI Tel. 0039 / 81 / 61 18 00
BARY-LENGER, A.	Centre Wallon du Bois Consortium AVSA rue de la Converserie, 44 B - 6900 SAINT-HUBERT Tel 061 / 61 15 48 or 061 / 61 15 55
BEENACKERS, A.A.C.M.	Twente University of Technology P.O.Box 217 NL - ENSCHEDE Tel. 053 / 89 30 37
BICKLE, R.	John Brown Engineers + Constructor Ltd Eastbourne Terrace GB - LONDON W2 6LE Tel. 01 / 262 8080 Telex 263521
BRIDGWATER, A.V.	University of Aston Dept. of Chemical Engineering GB - BIRMINGHAM B4 7ET Tel. 021 / 359 3611
BRUZZO, V.	CESEN Via Serra, 6 I - 16122 GENOVA Tel. 0039 / 10 / 58 62 41 Telex 27008 ansimp
CARRE, J.	Biomass Manager Commissariat Energie Solaire 208, rue R. Losserand F - PARIS 14 Tel. 533 00 05 Telex comes 203712 f
CASADEVALL, E.	ENSCP - Lab. de Chimie bio-organique et organique Physique. 11, rue Pierre et Marie Curie F - 75231 PARIS CEDEX 05 Tel. 336 25 25 - ext. 3836

CASERTA, G.	CNEN - Casaccia Via Anguillarese Km 1 + 300 I - 00060 ROMA Tel. 06 / 694 83 538
CHARTIER, P.	Project Leader INRA - Bioclimatologie route de Saint Cyr F - 78000 VERSAILLES Tel. 3 / 950 75 22
CHRYSOSTOME, G.	CREUSOT-LOIRE Division Energie B.P. 31 F - 71208 LE CREUSOT Tel. 33 / 85 / 55 80 80
COURTINAT, M.	Creusot Loire Entreprises 33, Quai Gallieni F - 92150 SURESNES Tel . 772 12 12 Telex 610062 f
DAVIES, R.J.	Foster Wheeler Power Products Greater London House Hampstead Road GB - LONDON NW1 Tel. 01 / 388 1212 x 280 Telex 263984
DEBBAUT, P.	AVSA - Association pour la valorisation des Vallées de la Sûre et de l'Attert Heinstert 65c B - 6722 NOBRESSART Tel. 063 / 21 38 24
DENNING, C.	Wellman Engineering Corp. Ltd Roberts House Cornwall Road GB - SMETHWICK, B66 2 JU Tel. 021 / 558 3151 Telex 337598
DERORE, J.	Société Creusot-Loire Division Energie 15, rue Pasquier F - 75383 PARIS CEDEX 08 Tel. 33 / 1 / 268 15 15 Telex motoy 650309 f
DE SANTI, M.	Regione Toscana Settore Energia Via S. Gallo 34/a I - 50100 FIRENZE Tel. 055 / 48 00 89 / 47 52 64 Telex 572508 retosc i

DIETRICH, R.

PLE - Nuclear Research Centre
517 Jülich
Aachenerstr. 74
D - 5162 NIEDERZIER KTN
Tel. 02461 / 61 / 4809

DUBOIS, P.

Laboratoires de Marcoussis
Route de Nozay
F - MARCOUSSIS 91
Tel. 449 12 00

EDWARDS, A.

Wellman Mechanical Engineering Ltd
Roberts House
Cornwall Road
GB - SMETHWICK, West Midlands, B66 2LB
Tel. 021 / 558 3151
Telex 337598

FIORITO, G.

CESEN
Via Serra 6
I - 16122 GENOVA
Tel. 0039 / 10 / 58 62 41
Telex 270008 ansimp

GEHRMANN, J.

KFA-Jülich
Abt. PLE
P.O.Box
D - 5170 JUELICH
Tel. 02461 / 61 48 52

GIBBS, D.F.

Int. Research and Development Ltd
Fossway
GB - NEWCASTLE UPON TYNE, NE6 2YD
Tel. 0632 / 650 451
Telex 53 7086

GILLIN, O.

Centre Wallon du Bois
44-46, rue de la Converserie
B - 6900 SAINT HUBERT
Tel. 061 / 61 15 55

GLYNN, P.

National Board for Science and Technology
Shelbourne Road
IRL - DUBLIN 4
Tel. 68 33 11

GRASSI, G.

Commission of the European Communities
Directorate General XII
200, rue de la Loi
B - 1049 BRUSSELS
Tel. 735 00 40 x 6801

GUDIN, C.

Dept. Biologie - Serv. de Radioagronomie
Centre d'Etudes Nucléaires Cadarache
B.P. 1
F - 13115 SAINT PAUL LEZ DURANCE
Tel. 42 / 25 71 05 or 25 70 88
Telex ceaca 440678 f

JANESCH, R.

Fritz Werner GmbH
Industriestrasse
D - 6222 GEISENHEIM
Tel. 06722 / 5011
Telex 042134

JASTER, K.

Fritz Werner Industrieausrüstungen GmbH
Industriestrasse
D - 6222 GEISENHEIM
Tel. 06722 / 5011
Telex 042134

KIMUS, J.

Centre Wallon du Bois
44-46, rue de la Converserie
B - 6900 SAINT HUBERT
Tel. 061 / 61 15 55 or 61 15 48

KING, G.H.

Energy Technology Support Unit
B 156, AERE Harwell
GB - DIDCOT, Oxon OX11 ORA
Tel. 0235 / 83 46 21
Telex 83135

KYRITSIS, S.

Professor
Agricultural College of Athens
Iera Odos 75
GR - ATHENS
Tel. 01 / 346 00 25 or 591 63 36

LICATA, R.

Vicolo della Palomba 19
I - ROMA
Tel. 06 / 65 66 902

LINDNER, C.

Lurgi Kohle und Mineralöltechnik GmbH
Bockenheimerstrasse 42
D - 6000 FRANKFURT/M
Tel 0611 / 711 95 29
Telex 41236-330 lg d

LINNEBORN, J.

Fritz Werner GmbH
Industriestrasse
D - 6222 GEISENHEIM
06722 / 5011
Telex 042134

LONG, G.

ETSU
B 156 AERE Harwell
GB - DIDCOT, Oxon
Tel. 0235 / 83 46 21
Telex 83135

MASSON, H.

Free University of Brussels
Centre for Ind. Research (AVSA Consortium)
50, avenue F.D. Roosevelt
B - 1050 BRUSSELS
Tel. 02 / 649 00 30 x 2942 or 2916

MISSONI, G.	AGIP Nucleare Via del Giorgione 59 I - 00143 - ROMA Tel. 06 / 59 00 93 45 Telex 610082 enidro i
MOLLE, J.F.	CEMAGREF Parc de Tourvoie F - 92160 ANTONY Tel. 1 / 666 21 09
MONHOVAL, L.	Centre Wallon du Bois Consortium AVSA 44-46, rue de la Converserie B - 6900 SAINT HUBERT Tel. 061 / 61 15 48 or 61 15 55
MOSS, G.	Esso Research Centre GB - ABINGDON, Oxon OX13 6AE Tel. 21 600 (office)
NAVEAU, H.	Université Catholique de Louvain Unit of Bioengineering 1, Place Croix du Sud, Bte 9 B - 1348 LOUVAIN LA NEUVE Tel. 010 / 41 81 81 x 3646 Telex 59037 ucl b
NICOLAY, D.	Commission of the European Communities Directorate General Information Market and Innovation Jean Monnet Building B4/072, P.O.B. 1907 L - 2920 LUXEMBOURG Tel. 43011 x 2946 Telex 3423/3446 comeur lu
NOACK, D.	Univ. of Hambourg and Federal Research Centre for Forestry and Forest Products Leuschnerstrasse 91 D - 2050 HAMBURG Tel. 040 / 73 91 91 Telex 214732 uni hh d
PSYLLAKIS, N.	Directeur du Centre de Recherches Agricoles de Crète Agrokipion GR - CHANIA Tel. 0821 / 89345 or 89346
NOEL, R.	INIEX - Institut National des Industries Extractives 200, rue du Chéra B - 4000 LIEGE Tel. 041 / 52 71 50 Telex 41128

PALZ, W.	Commission of the European Communities Directorate General Research, Science and Development 200, rue de la Loi B - 1049 BRUSSELS Tel. 735 00 40
PIRRWITZ, D.	Commission of the European Communities Directorate General Research, Science and Development 200, rue de la Loi B - 1049 BRUSSELS Tel. 735 00 40
REGINSTER, J.	Services de Programmation de la Politique Scientifique 8, rue de la Science B - 1040 BRUXELLES Tel. 02 / 230 41 00 Telex proscient bru b 24501
REIMERT, R.	LURGI Kohle und Mineralöltechnik GmbH Bockenheimer Land Str. 42 D - 6000 FRANKFURT Tel. 0611 / 71 19 537 Telex 41236-330 lg d
ROBINSON, K.	Energy Division National Board for Science & Technology Shelbourne House, Shelbourne Road IRL - DUBLIN 4 Tel. 68 33 11 Telex 30327 mbst ei
ROUSSE, J.	Creusot-Loire Entreprises 33, Quai Gallieni F - 92150 SURESNES Tel. 772 12 12 Telex 610062 f
RUCQUOY, A.	Université Libre de Bruxelles Centre de Recherches Industrielles 50, av. F.D. Roosevelt - B.P. 165 B - 1050 BRUXELLES Tel. 02 / 649 74 83 Telex univlibre brux 23069
SFORZA, C.A.	Regione Toscana Settore Energia Via San Gallo 34/a I - 50100 FIRENZE Tel. 055 / 48 00 89 or 47 52 64 Telex 572508 retosc i

SIRONVAL, C.

Université de Liège
Département de Biologie
B22, Sart-Tilman
B - 4000 LIEGE
Te'- 56 18 26

SKELLEY, W.N.

Pollution Prevention
(Consultants) Ltd
Ashby House
64 Hion Street
GB - WALTON ON THAMES, Surrey
Tel. 0342 / 71 44 39

TOCCHELLA, A.

CMI
Ansaldo Group
Corso Perrone 23
I - 16100 GENOVA
Tel. 0039 / 10 / 44 12 61
Telex 270318

VAN DER BURGT, M.T.

Shell Internationale Petroleum Mij BV
Carel van Bylandtlaan 30
NL - DEN HAAG
Tel. 070 / 77 23 46
Telex 31005

VAN SWAAIJ, W.

Twente University of Technology
Sportlaan 60
NL - 7581 BZ LOSSER
Tel. 05423 / 2677

WAGENER, K.

Technical University Aachen
Dept. of Biophysics
c/o Postfach 1913
D - 5170 JUELICH
Tel. 0049 / 2461 / 61 67 11

WILSON, H.

Foster Wheeler Power Products
Greater London House
Hampstead Road
GB - LONDON, NW1 7QN
Tel. 01 / 388 12 12

GPSR Compliance

The European Union's (EU) General Product Safety Regulation (GPSR) is a set of rules that requires consumer products to be safe and our obligations to ensure this.

If you have any concerns about our products, you can contact us on ProductSafety@springernature.com

In case Publisher is established outside the EU, the EU authorized representative is:

Springer Nature Customer Service Center GmbH
Europaplatz 3
69115 Heidelberg, Germany

Batch number: 10614746

Printed by Printforce, the Netherlands